Lecture Notes in Energy   2

For further volumes:
http://www.springer.com/series/8874

Michael R. Greenberg

Nuclear Waste Management, Nuclear Power, and Energy Choices

Public Preferences, Perceptions, and Trust

Michael R. Greenberg
Rutgers University
New Brunswick,
NJ, USA

ISBN 978-1-4471-4230-0   ISBN 978-1-4471-4231-7 (eBook)
DOI 10.1007/978-1-4471-4231-7
Springer London Heidelberg New York Dordrecht

Library of Congress Control Number: 2012945293

Printed on acid-free paper

Springer is part of Springer Science+Business Media (www.springer.com)

To my wife Gwendolyn Greenberg who has been listening to me talk about nuclear power plants, nuclear waste management, and chemical weapons stockpile sites for almost 40 years. Thanks for your patience and encouragement.

Foreword

For nearly 25 years the author of this book, Michael Greenberg, has devoted a significant portion of his prodigious research work to questions related to how public perceptions (particularly perceptions of risk) shape the policies, economics, and technology evolution/deployment of the nation's efforts to address protectively the wastes and other residuals of the nation's nuclear activities, both those associated with nuclear power and national defense. He has linked his work on nuclear issues to his own frameworks for also understanding how to assess and conduct other types of activity about which locally and nationwide US citizens often have diverse views ranging from ambivalent to strident—chemical weapons disposition, homeland security policies, and all manner of locally unwanted land uses (LULU's).

For 17 of those years, Greenberg's work on nuclear waste questions has been conducted within the context of a unique multi-university organization, the Consortium for Risk Evaluation with Stakeholder Participation (CRESP). CRESP was forged to respond to a 1994 National Academy of Sciences study that suggested a university-based independent organization was needed to help the newly formed Department of Energy Office of Environmental Management (DOE-EM) more effectively link its waste management, remediation, and compliance activities to both risk-informed priorities and risk-informed outcomes. CRESP won the initial competition to "be" that independent institution working to advise DOE-EM and has continued under several different forms and formats with essentially the same mission—although its work is no longer limited only to environmental management challenges from the nuclear defense legacy since it now formally also addresses waste management's parallel problems associated with residuals from civilian nuclear power. The organization has evolved from being primarily the merged activities of two research universities to being an organization, now led by Vanderbilt University's Department of Civil and Environmental Engineering that draws upon premier researchers and their laboratories at seven other major research universities. And it is always poised formally or temporarily to add types and areas of research capability to address whatever skill set is needed to allow the organization to provide the review, research, strategic assessment, or educational activities most needed by DOE-EM.

Within CRESP, Greenberg's own skill effectively to develop and carry out both public surveys and economic analyses has been performed in the closest possible proximity to and collaboration with applied academic research skills and work far different from his own: research, review, and strategic work in environmental engineering, chemical engineering, hydrology, ecology, nuclear engineering, materials science, community and occupational medicine and epidemiology, project management, biostatistics, soil chemistry, etc. Greenberg has made the most of the opportunities afforded by the fact that CRESP's unique agglomeration of academic people are able to give long-term attention to the complex changing social and technological phenomena that have evolved in the nation's struggle to achieve protective nuclear waste management. As both an analyst of those efforts and a commentator proposing different ways of performing them, Greenberg has pursued both his own set of research agenda and continuously collaborated with all of the other CRESP people in a diverse set of multiauthored CRESP work products. Probably no better example of this latter phenomenon exists than what is seen in the collaborative process which emerged under Greenberg's leadership the CRESP-organized and supported Vanderbilt University Press 2009 *Handbook on Nuclear Materials, Energy and Waste Management*. Furthermore, through CRESP, Greenberg has educated a cadre of diverse senior scientists and engineers in the practical implementation of stakeholder engagement and risk communication to have a more effective dialogue with the full range of interested and responsible stakeholders, from local communities to DOE managers, regulators, and oversight organizations.

As any careful reader of this book will soon recognize, Greenberg has seen his own distinctive work as providing a uniquely coordinated but yet evolving picture of where the nation as a whole, and the several clearly distinct geographical and ideational parts of it, has been and is today. He continues to clarify this, in the midst of what he so artfully describes as the persistently "mixed policy messages" (see, e.g., Chap. 1) sent by all manner of nuclear advocates and detractors, as well as the ever evolving and sometimes rapidly changing directions of the federal government. He wants to understand the diverse commitments and attitudes toward nuclear waste, how they evolve, and whether the materials to be managed were the products of weapons production, electrical power, medical applications, etc.

Hence, for example, as many CRESP people were carrying out research into projects seeking major shifts in technological or scientific assumptions and applications or discovering that the fundamental physics, chemistry, or biology associated with nuclear wastes and its proper placement in the environment opened- or closed-proposed remedial alternatives, Greenberg was focused on a parallel but different task. He was documenting the implications of precisely those same changes for the economic regions near nuclear facilities and how those activities created patterns of both economic growth and/or dependence.

Similarly his survey work was revealing parallel developments in how those project shifts were being perceived. And he made an exceedingly important finding about those who live near nuclear facilities—that familiarity with nuclear activity and the people and organizations who conducted it typically lead those nearby to perceive, in terms of safety, that nuclear facility activity at the neighboring site has been competently performed and is, with qualifications, acceptable to them. These neighbors simply have higher levels of trust in nuclear waste management and all things nuclear than does the general population. And Greenberg through those longitudinal studies tracked over time the rise and sometime the deterioration of trust in definable aspects of diverse waste management activities. On-site work is more trusted than transportation of waste, for example.

Over those 17 years, and particularly since 2004, the actual achievements of the projects intended to bring nuclear waste under effective management and control have rarely matched the expectations of their implementers, regulators, or those who provided funds to them. Acrimony has been persistent. Success in clarifying and implementing either new locations for or approved physical–chemical forms and geologic settings for the final disposition of waste was exceedingly rare. Some smaller DOE sites have sent their waste to the big ones—but always with difficulty and against some resistance.

Yet while the nation was fruitlessly trying to overcome having constructed a "Fuel cycle to Nowhere,"[1] relative calm has existed between those charged with the ongoing conduct or regulation of the very long-term efforts to complete the major waste management they could actually implement even when they did not know where the waste was ultimately going. And correspondingly, the unique social structures (site specific advisory boards) established by the Department of Energy both nationally and locally to mediate and communicate between the site professionals and the lay populations in the places where nuclear facilities had been constructed has worked surprisingly effectively.

Indeed, Greenberg's work has helped inspire and focus CRESP's own very successful efforts to make informed consent achieved through active public participation in completely transparent processes the hallmark of its own efforts actually to resolve seemingly intractable controversies into which the Department and its stakeholders have gotten themselves. Notably, when DOE and the people and state of Alaska could agree on almost nothing about how to "close" Amchitka, the Aleutian Island site where the USA had conducted three major underground nuclear tests, CRESP actually carried out an expedition and a major testing program to clarify whether radionuclides from those test sites had leaked into the maritime environment where both native subsistence and commercial fishing might be at risk. Resolution of the Amchitka closure challenge was accomplished through stakeholder engagement by CRESP throughout the program—from initial definition of the needed science plan, to Aleut participants on the sampling expedition, to extensive dialogue about the results of our findings through a series of public

---

[1] From the title *Fuel Cycle to Nowhere: U.S. Law and Policy on Nuclear Waste* of the Vanderbilt University Press book authored by CRESP researchers Richard Burleson Stewart and Jane Bloom Stewart (Nashville, 2011).

meetings in Alaska (that included visits to Aleut communities). A similar philosophy of stakeholder engagement has permeated the full range of CRESP activities at individual DOE sites, ranging from developing new regulatory approaches to reaching end-states, to understanding health impacts of mercury contamination in fish, and to development of fundamental technical understanding of waste form performance.[2] CRESP researchers (including the authors of this foreword) have been asked to transfer this learning and experience about the meaning of informed consent to the issues of reorganizing nuclear waste management disposition when most recently considered by the Blue Ribbon Commission on America's Nuclear Future.

As the years passed and the "dread" of another Chernobyl or Three Mile Island receded only slowly, the fact that nuclear power and nuclear waste management suffered no major catastrophes and the periodic fears of the consequences of media-reported risky behavior never resulted in actual failure with fatalities brought with it an increase in public acceptance of things nuclear. As we see in his book, Greenberg's longitudinal analyses adroitly tracked that evolution and provided insight into it. After the nation walked away from its $11B investment in a permanent geological repository at Yucca, it set up the Blue Ribbon Commission on America's Nuclear Future. When that Commission began documenting how the failure to secure informed consent "explained" why Yucca had been abandoned (a process in which CRESP researchers directly participated in presentations to the Commission), the Commission seemed—if temporarily—to be succeeding in shaping positively the national discussion. And, as Greenberg clearly indicates, the debate over whether nuclear is an antidote to anthropogenic climate change continued to attract support or criticism from the usual suspects, but the viability of nuclear energy as part of the solution seemed to be gaining.

As the end of 2010 approached and we as the coprincipal investigators of CRESP were charged with proposing to DOE what would be the most important contributions a multidisciplinary group of experts could make to accelerating protective nuclear waste cleanup, we asked: Were we really going to learn much more from the Greenberg survey work? Was CRESP's work on the "harder" sciences and technical pieces going to continue to be important in the contest of monumental technology choice and implementation challenges, accompanied by massive cost overruns, faced by the Office of Environmental Management? Our clear sense was "Yes!," that while not always clearly recognized, the insights gained from Greenberg's risk perception and risk communications research are critically important to developing constructive pathways forward.

And then, of course, came Fukushima and with it the reality that the uncertainties about nuclear energy were for most of the population very close to the surface and the impediments to building a national consensus about nuclear

---

[2] For additional examples and elaboration, see the book edited by CRESP researcher Joanna Burger, *Stakeholders and Scientists: Achieving Implementable Solutions to Energy and Environmental Issues,* New York, Springer, 2011.

power and waste management were again exposed clearly. As Greenberg clearly indicates, the power and specificity of the 2011 CRESP study in informing us about how Fukushima has played and will play in public perceptions of nuclear energy is largely dependent on what we had learned in the CRESP longitudinal studies that preceded it. There is no point in our explaining what Fukushima means since Greenberg does it so well in both Chaps. 5 and 6. But we note that this entire book is made possible by the existence of a longitudinal data that precedes the Japanese accident. And we are enthusiastic that our colleague saw the opportunity it presented to link his own specialty to the broader book-length understanding he provides here. The very last paragraph of the book does, however, remind us again of something that we continually worry about—that exclusive focus on the technological aspects of major societal challenges prevents successful resolution, because the social dimension of these challenges is a critical component that must be researched, understood, and be a part of any durable solution:

> Risk preferences and perceptions are typically referred to as "soft" issues, which I think is the wrong message to community representatives and their staff. Whatever the label, these issues preferences and perceptions do polarize people, cause fear, and undermine confidence in government and business that could linger and undermine genuine efforts to reduce risk. This means that elements of DOE, NRC, and EPA that need to rebuild trust in the short term and maintain it for generations must be strengthened, because proactive actions are more successful than reactive ones (Fischhoff 1995). I urge the government agencies to establish continuous education programs for managers and others who are likely to interact with the public and their representatives. This would be facilitated by the appointment of advisory boards with expertise in psychology, communication, sociology, planning and economics, and providing information for media (Greenberg et al. 2009; Environmental Health Center 2001; Lofstedt 2005).

We believe that CRESP, and we as its leaders, have understood this paragraph. It drives us to persistent efforts to improve policy assessment, so we will eventually send fewer mixed messages. And there cannot be protective disposition of the wastes while we still await determination of where the wastes should go and under what circumstances. And we are poignantly aware that inadequate understanding of technology readiness and its proper place in facility design remains a huge impediment to both meeting regulatory milestones and, as importantly, achieving publicly protective waste management results. CRESP has major contributions to make to each of these types of activity. But we have at times had to fight hard for the "place" in the overall CRESP effort of the work that Michael Greenberg does—and are proud that this book vindicates again our confidence in him and in what he brings to the full table of the resources needed to solve the nuclear waste management conundrum.

We also offer our sincere thanks to the series of perceptive Assistant Secretaries and senior managers of the Department of Energy Office of Environmental Management that have provided continuous support for the CRESP mission, including the work by Greenberg.

Nashville, TN, USA   David S. Kosson
                                                                                                                        Charles W. Power

Preface

December 9, 1953 was a special day for me, but I did not know it when I woke up. I was 10 years old, and mostly what I thought about was baseball (the Yankees had won the World Series again beating Brooklyn, as usual). School was something to be endured, while I prepared to be an outfielder for the Yankees. When I got to school on the 9th, I learned from my teacher, Miss Ducey, that President Eisenhower had spoken about atoms for peace on the 8th at the United Nations. Even though I was a young boy, I had visited the United Nations, so I created an image of the scene. I also knew who Ike was. My mother told me that she voted for him, which was a shock because I thought that my family always voted for Democrats.

That idea of nuclear energy really interested me. I did not know what an atom was. I did know that we dropped atom bombs on Japan and had been told that helped end the war, which was good because according to my dad two of my parents' friends would have been killed invading Japan. Eisenhower through some scientific miracle was going to take the energy of a nuclear weapon and turn it into cheap energy. Who was I to disagree with the President's dream? Miss Ducey told us that not only would we have power from nuclear atoms, but our nation has tremendous coal reserves, and our *Weekly Reader* had reported finding tremendous new oil reserves in the Middle East. In essence, we were set with respect to energy or at least that is the lesson I learned in 1953.

My interest in atoms and nuclear energy did not stop after President Eisenhower's speech. To the amazement of my parents and Miss Ducey, who already had me branded as needing improvement in "conduct" and "handwriting," and had hit me with a yardstick for not paying attention when she was teaching fractions, I insisted on going to the library and carrying home books so that I could read about atoms and chemicals. These are my first recollections about nuclear energy and weapons.

Once I figured out what an atom was, it was not long before I was bothering everyone I knew to explain how a nuclear weapon was made, how a weapon was different from a power plant, and what happened to all the nuclear waste: my uncle Sol called it the nuclear "trash." Unfortunately, I did not find the person I needed until I met Mr. Brown, my science teacher, who could explain the basic theory to me. It also helped that my sister Andrea was 4 years older, and I could borrow her chemistry and physics books.

I never played for the Yankees (stopped playing after college), nor did I become a chemical engineer because my chemistry teachers were afraid that I would set the lab on fire. So, the compromise was that I would major in mathematics and study science and history as secondary subjects. All through the 1950s and 1960s, I read every article I could about energy, especially nuclear energy.

In the 1970s, while still at Columbia University, I got a chance to witness my first of many risk communication failures involving nuclear energy when Columbia University decided to build a Triga Mark II nuclear reactor in the basement of the Engineering School without briefing the community. That failure led me to wonder about the wisdom of putting nuclear facilities in densely populated areas, especially without building a community constituency for the idea.

Two years later, my colleague Donald Krueckeberg and I were hired by Public Service Electric and Gas of New Jersey as part of their efforts to site a nuclear power plant about 10 miles north of Philadelphia, PA and about 4 miles south of Trenton, NJ at a place called Newbold Island. After completing our work, which consisted of land use and population studies, we faced 5 days of interrogation by 10–12 lawyers; none were in favor of the project. This experience made me recognize that emotions dominated cognitive processes in public settings.

The most riveting part of the hearing was the witness who got up, beat her fists on her chest, and declared in a decibel level that got everyone's attention that she was not going to allow the greedy utility to give her children cancer. In contrast to her outburst, the chief witness for the utility dressed in a white shirt, tie, and gray suit had to admit that he had never visited the site nor talked with any resident of the area. He had to admit that he did not know if the site was urban, nor how many people drove their cars through the exclusion zone. The closest he got to the site was a helicopter ride over the island. It got worse for him and the company when I testified to what we had learned about the area, which was that population growth had been converging toward the site.

The license was ultimately denied, and then the AEC took our work and used it as cumulative population density guidelines out to 50 miles for siting new nuclear power plants. We had no idea that they intended to use our work in that way, especially because they took our exact numbers and never bothered to ask about key land use and population growth theories. To be fair, they did build a siting group headed by one of our former students who provided badly needed expertise.

A few years later, Don Krueckeberg and I worked for the Nuclear Regulatory Commission to try to understand why people were moving into towns that hosted nuclear power plants instead of avoiding them. After some initial face-to-face interviews, we learned that not everyone is afraid of nuclear energy and that some people loved the idea of living in a nuclear power plant town, because the company paid an enormous portion of the town taxes. I will never forget the gentleman who could see the plant from his house telling me that he was more likely to be killed driving to and from work and that he paid almost no property taxes, and his children went to the best schools and had free violin lessons. He picked the town.

By the mid-1980s, it was pretty clear that I was going to continue to work on siting, population, and land use-related issues. Along with economic impact analysis and principles of risk analysis, those subjects are the major areas I have covered for the Consortium for Risk Evaluation with Stakeholder Participation (CRESP) for 15 years. Centered at Vanderbilt University, CRESP has scientists, engineers, and social scientists from the University of Arizona, Georgia Tech, Howard, New York University, Oregon State, Robert Wood Johnson Medical School, Rutgers, and Wisconsin, and CRESP has been funded by DOE. As the name suggests, CRESP understands the importance of the public, especially the public that lives near existing sites. In 2005, 2008, 2009, 2010, and 2011, it funded population surveys, the objectives of which have been to fill a gaping hole in what we know about public perception and preferences about nuclear waste management, energy, and technology. We, principally me, write the questions, not the DOE.

Given the polarization around nuclear power that we hear about, it is only fair that I state my views so that the reader need not waste time trying to guess which side of the fence I sit on. I uncomfortably sit on the fence. With regard to nuclear power, I look at the entire fuel cycle, defense waste, and other energy options, not just nuclear power plants. My record as briefly described above is having worked on studies that have tended to make it more difficult to site new nuclear facilities and to populate communities that have them. I started with great curiosity about nuclear energy and weapons, not with a conviction that I knew what the country should do. My frustration is that we still do not have a clear energy and transportation plan that has been articulated to the public. How are we supposed to evaluate nuclear energy, unless we have some idea about the country's ideas about coal, gas, and the other options?

Fukushima, like TMI and Chernobyl, complicates the policy formation process. The images were painful, and the easy policy for some is no new plants and close existing ones as soon as possible. Whether President Eisenhower's dream has become a nightmare is a bona fide question to ask. Personally, I would like to see the country engage in a reasoned debate about the fate of commercial nuclear power rather than be forced into more coal and gas plants without the benefit of a cognitive process that somehow faces the reality that we need to continue to plan investment in energy sources to match the growing electricity consumption. In this context, there is much talk about energy conservation and a sustainable future, but there is much more talk than there is action. If we are serious about reducing demand, then as stated above I need to see a real plan and program, not just talk but appeal to the public.

But while the combatants in the nuclear power debate joust and the media writes stories about the jousts, the future of nuclear power does not excuse us from facing the reality of accumulated defense waste siting at facilities that need to be treated for the indefinite future, as does the commercial nuclear waste. We need a plan to deal with two streams of nuclear waste or a hybrid of the two no matter what we decide about nuclear power plants. This daunting challenge is the focus of this book as seen through the eyes of thousands of US residents who we surveyed five times from 2005 through 2011. Their preferences and perceptions tell us much more about their emotions, images, trust, and personal history than about their knowledge of nuclear waste or energy, but no matter the basis of their views, it would be incomprehensibly foolish to ignore them as irrational and not build a relationship that will need to exist for the indefinite future while the already accumulated waste stockpile degrades to less harmful forms or is reused.

New Brunswick, NJ   Michael R. Greenberg

Acknowledgments

Beginning in the mid-1970s, I worked closely with Donald Krueckeberg, my colleague at Rutgers for over 30 years, on nuclear power plant siting issues for the Atomic Energy Commission and the Nuclear Regulatory Commission (NRC). Later, Don and I worked on nuclear power issues for the NRC with Michael Kaltman, William Metz, and Kenneth Pearlman. We conducted research and testified before government review panels.

Since 1995, I have been a part of the Consortium for Risk Evaluation with Stakeholder Participation (CRESP). This multi-university group has allowed me to meet and work closely with dozens of eminent scholars in engineering, physics, chemistry, hydrology, ecology, medicine, public health, law, and ethics from across the USA. Charles Powers, David Kosson, Bernard Goldstein, Joanna Burger, Michael Gochfeld, Henry Mayer, Karen Lowrie, and Heather Truelove have been members of the CRESP team and worked on several of the surveys described in this book. Without their ongoing encouragement and participation, these surveys probably would have stopped after the first two. Every year, a group of CRESP members would try to define the key questions to investigate during the following year, and then I would design a survey to answer those questions. I would like to thank Dr. Marc Weiner for helping me rewrite some of the questions so that they would be easier to understand and for engaging Abt SRBI to carry out the most recent surveys. I thank Jennifer Rovito for turning my crude ideas for figures into the maps and charts in this book. I appreciate the reviewers and editors who provided constructive suggestions of our published papers from the 2005, 2008, 2009, and 2010 surveys that are summarized in Chap. 5.

Resources for survey and staff support for the CRESP surveys were provided under three cooperative agreements with the US Department of Energy. The most recent is Cooperative Agreement Number DE-FC01-06EW07053 entitled The Consortium for Risk Evaluation with Stakeholder Participation III awarded to Vanderbilt University. The opinions, findings, conclusions, or recommendations expressed herein are those of the author and do not necessarily represent the views of the Department of Energy or Vanderbilt University, or any of the people acknowledged. This report was prepared as an account of work sponsored by an

Agency of the US Government. Neither the US Government nor any agency thereof, nor any of their employees, makes any warranty, express or implied, or assumes any legal liability or responsibility for the accuracy, completeness, or usefulness of any information, apparatus, product, or process disclosed, or represents that its use would not infringe privately owned rights.

All of the CRESP surveys in this book were reviewed and approved by the Rutgers University IRB.

Contents

List of Figures

List of Tables

Chapter 1
Managing the Nuclear Legacies

Abstract The USA has accumulated a great deal of nuclear and chemical wastes as a result of the production of nuclear weapons and nuclear power. An unambiguous path to managing these legacy wastes, which continue to increase, has been blocked by science and engineering uncertainties, nuclear proliferation concerns, high costs, and compounded by the absence of a comprehensive national government policy framework and a clear public consensus about the issues. The public hears inconsistent and often contradictory messages from advocacy groups and the media, as well as from senior government elected officials and staff who have the responsibility to manage the wastes. In order to understand public preferences and perceptions about new nuclear missions at existing major DOE sites and public preferences for alternative electric energy fuel sources, CRESP surveyed US residents, disproportionately near six major DOE sites, in 2005, 2008, 2009, and 2010. In 2011, after the Fukushima events, another survey was conducted to determine the impacts of the Japanese events on the preferences and perceptions observed in the four earlier surveys.

1.1 Introduction

Safely and efficiently managing the ongoing ecological and human health legacy created by nuclear weapons and nuclear power plants are two vexing interconnected challenges, especially so in the USA, the UK, France, Russia, and other nations that have had both nuclear weapons and nuclear power plants. This chapter introduces the challenges. It begins by summarizing five intertwined issues faced by nuclear waste managers: science and technology, proliferation, economic costs and benefits, multilayered policy puzzles, and public participation in areas with nuclear facilities. Part two connects the five issues to mixed messages to the public about nuclear science, power plants, and especially waste management. The mixed messages undermine public trust and condition the population to react strongly to events rather than trust their leaders to pursue the most effective path.

M.R. Greenberg, *Nuclear Waste Management, Nuclear Power and Energy Choices*,
Lecture Notes in Energy 2, DOI 10.1007/978-1-4471-4231-7_1,
© Springer-Verlag London 2013

Part three summarizes Chaps. 2–6 beginning with the construction of the US nuclear industrial complex to build nuclear weapons and ending with the need to build ongoing meaningful relationships with those who represent communities near the legacy nuclear facilities.

1.1.1 Science and Technology

The scientific and engineering challenges associated with nuclear weapons, power plants, and waste management are unprecedented. It is true that much of basic science was known and some technology existed before the USA chose to build nuclear weapons. But the scale-up from ideas and bench science to building weapons, power plants, and managing their wastes has been a massive endeavor. Something that works in a laboratory and in a small test facility may not work in a full-scale operational one. Much of nuclear waste management technology, which includes not only nuclear wastes but also chemicals and mixes of nuclear materials and chemicals, has been developed or substantially refined during the last 3 decades. Some of the most brilliant engineers, physicists, chemists, and risk scientists have devoted their careers to making the weapons, designing nuclear power plants, and minimizing risk, and many equally talented scientists have worked on dismantling the facilities and managing the waste. Yet a simple set of scientific solutions for waste management has proven elusive.

The technological challenges are not only due to the need to manage the wastes as safely as possible within budget limitations, but the temporal scale of the challenge is daunting. The first Egyptian pyramids were built about 4,500 years ago. Scientists can calculate how long it takes for radioactive and chemical elements to decay to less harmful forms and hence how long the sites that retain the residuals need to be protected. Much of the waste will be less dangerous in a century, but the potential risks of some nuclear wastes last for tens of thousands of years. No waste management problem that I am familiar with requires management on a forever time scale. Even assuming stable political organizations, how can even the best minds develop organizational processes and technologies that eliminate the likelihood of serious exposures?

Not surprisingly, science fiction writers have embraced the subject, creating events that seem ridiculous but have attracted public viewing. For example, a television show during the late 1970s, *Space 1999*, which the author watched, was built around a place called Moonbase Alpha, which stored nuclear waste from the Earth. A malfunction associated with the nuclear waste caused an explosion sending the moon hurtling out of the solar system. Fast-forward several decades, and many children watch the Simpsons, a cartoon show that features three-eyed fish perhaps caused by the owner of the nuclear plant illegally disposing of nuclear waste. These media portrayals of nuclear waste may make experts laugh, but they plant a bad seed with the public about the safety of nuclear waste management. Accordingly, when we asked US residents about their image of

DOE's weapons sites, we fully expected some to identify three-eyed fish from the Homer Simpson show.

1.1.2 Nuclear Weapons Proliferation

The international implications of nuclear waste management are a provocative challenge. Two illustrations show why. Currently, agreements between the USA and Russia require that both sides dismantle thousands of nuclear weapons, remove the fissionable material, and convert it into mixed oxide fuel (MOX) to be burned in nuclear power plants or otherwise rendered inaccessible to those who would make nuclear weapons. The US nuclear weapons are dismantled at the Pantex Plant near Amarillo (see Chap. 2). The international agreement requires cooperation between the USA and Russia, between the US Department of Energy (DOE) and commercial operators of nuclear power plants, designers and operators of facilities to convert MOX into commercial fuel, and other federal agencies as well as the DOE to ship the new fuel to nuclear power plants, and to decide where the waste from this conversion will be held and retained in what form. Burning MOX fuel in a nuclear power plant is complicated, as are the other options.

The DOE's Global Nuclear Energy Partnership (GNEP) was an attempt by the USA, France, China, Russia, and Japan to become the world's nuclear fuel suppliers, which would reduce the possibility of the fuel and waste being diverted to a weapon, as well as increase the use of nuclear power and thereby reduce fossil fuel use. In September 2007, the nuclear power nations were joined by Australia, Bulgaria, Ghana, Hungary, Jordan, Kazakhstan, Lithuania, Poland, Romania, Slovenia, and Ukraine as signatories of GNEP. Italy, South Korea, Canada, Senegal, and the UK later signed on to this worldwide effort (Greenberg et al. 2009). However, GNEP has been essentially discontinued by the Obama administration, and the project has lost momentum, perhaps permanently.

Wagner et al. (1999) argued for a new nuclear fuel cycle that would consume the plutonium. Von Hippel (2010) concludes that nuclear energy could make a "significant contribution to the global electricity supply. Or it could be phased out..." He ends by stating "If the spread of nuclear energy cannot be decoupled from the spread of nuclear weapons, it should be phased out" (Von Hippel 2010, p. 4). These two views and the failure of GNEP underscore the connection between nuclear weapons, power, and waste management, and the challenge of finding workable solutions at the international scale.

1.1.3 Economic Costs and Benefits

The costs of stabilizing, remediating, and cleaning up the DOE weapons waste legacy have already surpassed $100 billion, and perhaps it will reach $400 billion

(Office of Environmental Management 1996; Committee on Commerce 2000; Top-to-Bottom 2002; US DOE 2006; Huizenga 2012). Efforts have been made to reduce cost, such as accelerating cleanup at smaller sites and being more efficient. Some of these have been implemented, but the high costs are a challenge for the DOE (US DOE 2006; Top-to-Bottom Review Team 2002).

High costs are not a problem for those who derive economic benefit from the projects. Large waste management facilities are among the most unpopular locally unwanted land uses (Lindell and Earle 1983). Yet, in graduate school, the author's adviser reminded him that what was black and brown gunk to some people was green cash to others. Some of the DOE waste management projects are already the most costly ones in the world. Chapter 2 describes the economic impact of these projects on several regions in the USA. A discontinuation of environmental management funds would cause an economic depression in these places (Greenberg et al. 2002, 2003). Indeed, some local officials want new nuclear waste missions. The obvious example is elected officials in New Mexico who negotiated over many years with DOE for the Waste Isolation Pilot Plant (WIPP) (see Chap. 2 for more detail). But New Mexico is not the only area that has expressed interest. US Representative Jean Schmidt, up for reelection, called for consideration of a waste storage site in her district, which includes the DOE's former Portsmouth Gaseous Diffusion Plant near Piketon, in southern Ohio (Wilkinson 2006). In addition, about 90 % of DOE's budget is spent on contractors who have a vested interest in continuing the work.

Given the Fukushima event, the cost of managing the waste from nuclear power plants has slipped down the list of considerations, but not for long. The decommissioning and used fuel management costs are a major issue with nuclear power plants. The immediate issue is paying for the new plants. The idea of investing billions of stockholder dollars in a nuclear power plant that might or might not encounter delays because of technology, permits, public opposition, and other elements led the federal government to provide loan guarantees for new plants, for example, over $8 billion for the Vogtle facilities on the Savannah River in Georgia. Will this program continue? Without it, I find it hard to believe that programs to build new nuclear power plants will continue.

In the book the *Future of Nuclear Power* (MIT 2003), the authors discuss cost, safety, waste, proliferation, and many of the other issues and focus on some key decisions, for example, a once-through fuel cycle for nuclear materials. Their early and later reports (MIT 2003; Deutsch et al. 2009) illustrate the complexity of the nuclear power issue and the key part that cost and benefit play in the decision-making process.

The Fukushima events put even greater pressure on owners, operators, and government to consider the wisdom of allowing new nuclear facilities or continuing the operation of existing ones in earthquake, tsunami, flood, hurricane, tornado, and ice storm prone areas. The cost of upgrading existing plants, protecting them against these hazards and terrorism versus closing them is a major issue.

1.1.4   Multilayered Policy Puzzles

Policy is the penultimate challenge in this five-part synopsis. Some of the challenge is obvious. The DOE, US Nuclear Regulatory Commission (NRC), and other responsible parties are mandated to carry out agreements that are hard to achieve because of technical limitations and high costs, and yet trying to change those agreements has proven as difficult as the effort to meet the goals in the agreements (Stewart and Stewart 2011). In Chap. 2, I briefly describe the key laws and how they connect to managing the nuclear legacy. A key red flag is the inability of the federal government to fulfill its obligations under the Nuclear Waste Policy Act that require it to take possession of and permanently dispose of used commercial fuel in a permanent repository. A \$25+ billion trust fund has been accumulated and not spent toward this purpose, leaving the utilities to retain the used fuel on site and file over 70 lawsuits to recover the money (Cawley 2010; see also Chap. 6). The disposition of used nuclear fuel is central in this debate. In February 2011, New York, Vermont, Connecticut, the Natural Resources Defense Council, and others challenged the Nuclear Regulatory Commission's decision that extends the time allowed for storing spent nuclear fuel from 30 to 60 years (NJ DEP 2011). The challengers won.

Stepping back from these immediate policy challenges, the absence of a comprehensive and comprehensible energy policy, including waste management, is a serious problem. Michael Graetz and Ian Sharpiro's (2005) *Death by a Thousand Cuts: The Fight Over Taxing Inherited Wealth* explains the torturous evolution of the US estate tax. Graetz's (2011) *The End of Energy* could just as easily have been called death by at least a hundred cuts. In this new book, Graetz analyzes more than four decades of US energy policy beginning with the Nixon Administration's policy response to the oil embargo instituted by OPEC, and carrying on the lost opportunity to develop alternative energy sources, through today's dilemmas about fossil fuels, global climate change, and national security. Graetz emphasizes the reliance on technology-based solutions, the political need to keep energy prices low. Through the lenses of someone who has been interested in the subject for as many years as Graetz, the only consistent policy message I always see is that we need cheap energy prices. Otherwise, the numerous laws, guidelines, and regulations seem to be uncoordinated and even contradictory.

Nowhere is this lack of policy clarity clearer in the waste management part of the nuclear energy life cycle. Former US Senator Peter Domenici from New Mexico has been one of the strongest spokespersons for a nuclear energy renaissance directly related to the US's economic future, and he ties waste management to success. In a 2009 speech, Domenici (2009) said:

> The United States lags in the development and deployment of new nuclear technologies. We are quickly falling behind and will forfeit our historical leadership on this issue if current trends continue. America stalled thinking about used nuclear fuel. Other countries have developed, or are in advanced stages of developing strategies to address waste and non-preparation concerns. We are stuck in policies of more than 30 years old. We should not advocate use of nuclear energy, unless we [are] willing to get serious about the nuclear waste problem. During the last 12 years, America has moved backwards, not forwards, and addressing this challenge. We must fully engage the public in this effort.

Richard and Jane Stewart's book *Fuel Cycle to Nowhere* (2011) details the legal nightmare of nuclear waste policy in the USA. Their book is painful to read because it underscores that the USA does not have a coherent set of policies with regard to nuclear waste management.

The best example is the use of Yucca Mountain as the deep geologic repository. Located about 90 miles north of Las Vegas in an area with few people, the DOE has spent over $10 billion on the project to build this repository about 1,000 ft below the surface. The estimated cost is now closer to $100 billion, and the DOE prepared a series of environmental impact statements for this selection. The need for a permanent repository is not immediate, but there is need for interim storage or final storage plan that can be agreed upon and implemented.

Perhaps the Yucca Mountain siting would have been more successful had an independent body not involved in day-to-day waste management issues had been in charge. Various attempts to change DOE's waste management responsibilities have been proposed. Milton Russell (1997) argued that DOE was being held hostage by local governments that depend upon DOE funds to support their economy. He called for "productive divorce" that would separate the cleanup mission from the economic transition ones. In 1996, Caputo (1996) argued that Congress should end DOE's self-regulation of nuclear weapons activity. He called for control by an independent body that would be accessible to citizens. Notably, the first recommendation of the President's Blue Ribbon Commission on America's Nuclear Future (2012) was to call for a single-purpose federal corporation to manage nuclear waste.

1.1.5 Public Participation in Areas with Nuclear Facilities

The public is last in this list of five, but not the least important portal into the challenges of nuclear waste management. In a democratic society, we need to know what everyone thinks about nuclear power, nuclear waste management, and the disposition of nuclear weapons and their by-products. But in the case of nuclear materials, with all due respect to stock holders and rate payers, the people who live near the sites and along the paths that lead to the sites are the most important because they could be disproportionately burdened. The kind of informed consent a physician would obtain from a patient before a surgical procedure or that I would ask from someone who was being interviewed for this book are not the goals here. The DOE, NRC, EPA, and other responsible parties are charged with managing legacy wastes, and they cannot give away their responsibility to the public, public representatives at the sites, or anyone else. Even though they cannot obtain signed formal consent, they certainly can and should try to inform the public and listen closely to its views. If a technology is risky, is controversial, and has imbedded social and cultural values, then a strong government effort is essential, especially after the events in Japan in 2011. At a minimum, a bona fide effort could eliminate the charge that the federal agencies are deceitful and lie to the public.

Even before Fukushima, waste management is not something that people like to associate with. The familiar expressions locally unwanted land use (LULU), not in my backyard (NIMBY), not in anybody's backyard (NIABY), not in my term of office (NIMTOO), and build absolutely nothing anywhere near (BANANA) testify to the power of public opposition to a variety of facilities including landfills, incinerators, waste storage, and most certainly nuclear facilities (Greenberg 2009). The opposition is to the transportation of waste materials, and to the smells, odors, and visual blight, and stigma that can reduce property values, as well as to human health and ecological hazards.

With regard to defense nuclear waste, NIMBYism was brushed aside. The nuclear weapons sites were chosen during a war or shortly thereafter; in some cases, people were removed from their land. But the DOE recognized that their massive environmental management investments could be jeopardized by stringent public opposition and has site specific advisory boards (SSABs) at many of its key sites. The SSABs provide input to the DOE about its proposed waste management projects, so the public is not entirely ignored. Public input is solicited whenever DOE has to prepare an environmental impact statement or environmental assessment. In reality, the biggest advantage the DOE sites have is distance from major population centers (see Chap. 2).

Nuclear power plant wastes are stored at the active sites, which are not located in cities. Indeed, efforts to site a nuclear power plant in New York City were defeated as was one proposed in the Delaware River 10 miles north of Philadelphia and 4 miles from Trenton, New Jersey. These plants were relocated to more "remote" sites, which in the author's experience translates to be about at least 25 miles from a major urban center (Greenberg et al. 1984, 1986). On-site waste storage was a nonissue when these location decisions were made. After the public saw images at Fukushima, the issue can no longer be ignored.

1.2 Showered with Mixed Messages

The public consistently hears, sees, and reads contradictory assertions about nuclear waste, nuclear weapon decommissioning, and nuclear power. In the absence of a clearly articulated and persuasive set of policy decisions, the public is left to weigh the marketplace of contradictions about the hazardousness of nuclear energy and waste management options, transportation related risk, the value of nuclear power plants as an alternative to fossil fuels that cause global climate change, and many others.

Trying to provide credible information to the public is a challenge. Jenkins-Smith and Silva (1998) studied how the public copes with the mass of inconsistent and contradictory data in a sample of 1,800 residents (2/3 from New Mexico and the remainder for the USA as a whole). The public anticipates the positions spokespersons take about nuclear waste disposal, they favor research done by independent scientists, and in general they are skeptical about anyone seen as not neutral. In other words, well-articulated arguments grounded in good data do necessarily carry the day, and perceived independence is important.

While the public favors "objective" sources, even sources that are assumed to be independent of the energy industry present contradictory messages. For example, presumably, the *New York Times* is a credible media source. Yet, for example, Keller (2002), in a story in the *New York Times* with the title "nuclear nightmare," asserts that "experts on terrorism and proliferation agree on one thing: sooner or later, an attack will happen here, and the bottom of the front cover of the magazine ends with "How scared should we be?" The *New York Times* also published "Atomic Balm" (Gertner 2006), which makes a case for nuclear power and ends the front page by stating "But will it ever be … not scary?" The *New York Times* has published many other stories similar to these, and my question, aside from the clever choice of words, is what is the takeaway message to the public supposed to be? The only consistent message this writer finds is that we do not know what we should do about nuclear power. Needless to say, I am not picking on the *New York Times* because it along with only a few other newspapers presents interesting stories on the subject. Rather, my concern is that these media reports leave readers with little clarity.

It certainly does not help the public image when the DOE is criticized by Congress (Committee on Commerce 2000) with the provocative title "Incinerating cash." Despite many efforts, the DOE has not been able to implement a new business model that would deliver projects on time and close to projected costs. Arguably, given the complexity and newness of the technology, it is unreasonable to hold this technology to a private business model standard that is supposed to accomplish a task on time, on budget, and to technical specification (Pinto 1998; see also General Accounting Office 2001; Probst and Lowe 2000).

Another gap in public confidence about nuclear waste management was widened when on April 29, 2009, 17 members of the US Senate wrote to Secretary Energy Chu (Inhofe et al. 2009) questioning the removal of Yucca Mountain as a nuclear waste repository option. Paragraph 2 of the three-page letter begins as follows:

> Since the 1950s, 55 studies have been conducted by the NAS, in addition to numerous conducted in our National Labs and in international scientific bodies, as to the options and alternative to nuclear waste disposal. … Over \$7.7 billion has been spent researching Yucca Mountain as a potential repository site and neither the NAS, the NWTRB, nor any of our National Labs involved in conducting studies and evaluating data have concluded there is any evidence to disqualify Yucca Mountain as a repository. The scientific work resulted in a license application exceeding 8,600 pages, and was successfully docketed with Nuclear Regulatory Commission.

The authors of the letter concluded by posting seven questions for the Secretary to respond to; the first was: "What is the reason for your decision that Yucca Mountain is 'not an option.'"

Fortunately for all the parties involved, this interchange did not get much attention in the public media.

One level further up the hierarchy of power, the DOE itself sends mixed messages. In January 1995, Secretary of Energy Hazel O'Leary was clear about the role of environmental management in *Closing the Circle on the Splitting of the Atom* (OEM, DOE 1995, p vii).

> The United States built the world's first atomic bomb that helped win World War II and developed a nuclear arsenal to fight the Cold War. How we unleashed the fundamental power of the universe is one of the great stories of our era. Is a story of extraordinary challenges brilliantly met, a story of genius, teamwork, industry, and courage.
>
> We are now embarked on another great challenge of the new national priority: refocusing the commitment that built the most powerful weapons on Earth towards the widespread environmental and safety problems of thousands of contaminated sites across the land. We have a moral obligation to do no less, and we are committed to producing meaningful results. This is the honorable and challenging task of the department environmental management program.
>
> Although the war that gave us the atomic bomb a half-century ago, and the Cold War that followed is now over, the full story of the splitting of the atom is yet to be written.

In contrast, the role of environmental management in DOE's recent Strategic Plan is circumspect, at best (US Department of Energy 2011):

> Secretary Chu wrote: The Department of Energy plays an important and unique role in the U.S. science and technology community. The Department's missions and programs are designed to bring the best scientific minds and capabilities to bear on important problems. It is an integrator, bringing together diverse scientists and engineers from national laboratories, academia, and the private sector in multidisciplinary teams, striving to find solutions to the most complex and pressing challenges. This Strategic Plan lays out the Department's leadership role in transforming the energy economy through investments in research, development of new technologies, and deployment of innovative approaches.

The DOE Strategic Plan is organized into four distinct categories, representing the broad crosscutting and collaborative efforts taking place across the department's headquarters, site offices, and national laboratories. Those include:

- Catalyzing the timely, material, and efficient transformation of the nation's energy system and securing US leadership in clean energy technologies
- Maintaining a vibrant US effort in science and engineering as a cornerstone of our economic prosperity with clear leadership in strategic areas
- Enhancing nuclear security through defense, nonproliferation, and environmental efforts
- Establishing an operational and adaptable framework that combines the best wisdom of all department stakeholders to maximize mission success

Those who understand the DOE's role in environmental management can read between these lines to see environmental management. However, it certainly would not seem like a priority to a nonexpert journalist or member of the public. With an overall budget that quadrupled from $1.6 billion to over $6 billion between 1989 and 2000, this mission statement sends a mixed message about nuclear waste management.

Some might interpret this plan as part of a step to disconnect the DOE environmental management mission, or at least marginalize it. Indeed, previous proposals have been to send EM to the EPA, to privatize much of its mission, and the Blue Ribbon Panel (2012) suggests a new entity. I do not offer a position on any of these proposals in this book. I do, however, note that these are not messages that lead to building trust for the EM mission within DOE.

The US Nuclear Regulatory Commission (NRC) has had several confidence-shaking events. For example, in December 2011, a publically accessible debate broke out between Gregory Jaczko, the chairman, and all four commissioners. The essence of the debate is how rapidly should the NRC require actions by US utilities in response to the Fukushima events (Northey 2011). Irrespective of the merits of the charges and countercharges, and people part of the disagreement, many members of the public are likely to perceive this debate as government bureaucrats cannot be trusted to protect them. When people get mixed messages from people who are chosen to represent the technological elite of the country, the credibility of the organization suffers.

The final set of mixed message illustrations comes from recent US presidents, who presumably are focused on identifying the path that is best for the nation as a whole. I focus on two prominent issues of a nuclear power renaissance and the Yucca Mountain permanent repository. President George W. Bush was perceived as a strong supporter of nuclear power and the Yucca Mountain repository. He characterized nuclear power as "one of the safest, cleanest sources of power in the world." And he called for more plants, identifying them as a key part of his approach to reduce US greenhouse gas emissions. President Bush supported the idea of loan guarantees for utilities to build nuclear power plants and expedited licensing (Pegg 2005). With regard to the Yucca Mountain project, he agreed with DOE's recommendation to move ahead. Furthermore, the president strongly supported the GNEP program as an antinuclear proliferation policy (NEI 2002, 2003). Overall, in this writer's opinion, President Bush's positions appear to be consistent on these two issues.

President Obama's polices have been similar in some ways but not in others. He has continued to support nuclear energy, along with wind, solar, clean coal and natural—as clean energy sources (MSNBC 2010). With regard to nuclear power, he supported the granting of over \$8 billion to the Southern Company to construct the Vogtle nuclear plant in eastern Georgia. After the Fukushima events, President Obama continued his support for nuclear power, although acknowledging a down-side while assuring the public that the safety of all US nuclear plants would be reviewed (CNS News 2011; Wallenstein Lynn Yang 2011). Yet, with regard to the Yucca Mountain facility, he in essence stopped the project after DOE had submitted a massive environmental pact statement to EPA.

If we go back to President Clinton and to President Clinton's new book, *Back to Work* (Clinton 2011), the public hears another mixed message. President Clinton (2011, p. 351) states: "the one area where I disagree with the administration (Obama) is in the providing of large loan guarantees to nuclear power." While characterizing himself as not "instinctively antinuclear," the former president strongly advocates for solar. Furthermore, while the Yucca Mountain project moved forward during his administration, he notes that we have not solved the waste problem. During her presidential candidacy, candidate Hillary Clinton (Ball 2008) was quoted as saying that "Yucca Mountain will be off the table forever."

The inconsistencies in the positions of the last three American presidents started long before these three presidents. But even if we limit our views to these last three, what should a thoughtful American conclude from the messages? There is no consistent policy on either of these issues. Anecdotally, shortly before she passed away at the age of 96, my mother who was always interested in all sorts of public policy debates asked me about what was behind these mixed messages going back to the days of President Eisenhower. Her reaction to what she believed should be a science-based decision process was to be frustrated, lose trust in nuclear power science, and try to disconnect from the issue in favor of something that appeared to be more clearly articulated.

A national energy policy would address the life cycles of alternative energy sources beginning with obtaining fuel to final disposition and stewardship of waste products. It would be linked to laws, rules, and regulations regarding energy efficiency, economic incentives and disincentives, emissions of contaminants, transportation and storage alternatives, and how these are associated with national security and the nation's economic health, and with state and local, and private actions. It would include coal, natural gas, solar, wind, biofuels, hydropower, oil, and nuclear. In the case of nuclear, the policy would also have to explicitly deal with defense-related nuclear waste. All of these elements have been addressed in various places and various times during the last half-century. However, this author is hard pressed to identify a comprehensive and integrated set of documents that could serve as the basis for explaining options to the population.

If a plan existed, and the leadership consistently refereed to it and provided nuanced versions of it, then what Shrum (2001) called mainstreaming would occur (see also Shanahan 1995; Belsey 2008). That is, those who followed media reports of the policy would more likely have similar views, even among those who might, for a variety of reasons, be predisposed toward opposite views. Arguably, France and Austria illustrate nations with clear policies about increasing reliance on nuclear power, and these are reflected in the public surveys (see Chap. 3). When, however, there is no comprehensive and widely communicated plan, then preferences and perceptions become substantially conditioned by recent events such as Fukushima.

Without a set of coherent policies, we are vulnerable to what Spangler (1981) called "syndromes." He listed and described 73 syndromes, many of which will be familiar to readers. I cannot possibly due justice to his fascinating list, but here are three listed as technological progress syndrome: "gee-whiz" (scientific miracles will solve the issue), "small-is-beautiful" (technology is fine but needs to be small and simple), "thou-too-shall pass" (if we wait long enough opposition to technology will evaporate).

With regard to conditioning by recent events, we have had many events in the recent past, for example, the coal impoundments break in Kingston (TN) in 2008, the Deepwater horizon blowout and spill in 2010, and the Fukushima event in 2011. The first two arguably should condition the public to be more favorably disposed to nuclear power as an alternative energy source, and the third should lead to a call for less reliance on nuclear power. Yet, if the public perceives these events as signals of the inability of government officials to properly regulate and even manage energy technology, then the core federal agencies (DOE, NRC, EPA, DOI) will lose credibility. If the public blames these events on corporate greed, then they might be less trusting of utilities that own and manage nuclear and other plants, as well as those who manage the transport and waste management parts of the cycle. In other words, the experiences of encountering mixed messages from responsible parties increase the likelihood of dramatic changes in public preferences and perceptions. This is not to say that events should be ignored, but rather absent a well-articulated plan that has been part of popular culture, the public has no larger context from which to judge the severity of the hazard events. My expectation was that the Fukushima event would markedly condition public preferences and perceptions but that these would be tempered by the consistent media attention to global climate change. Chapter 5 shows that this expectation was found.

1.3 Organization of the Book

The annual surveys conducted by the Consortium for Risk Evaluation with Stakeholder Participation (CRESP) in 2005, 2008, 2009, 2010, and 2011 and featured in this book were conducted in order to address three questions that follow from the stream of mixed messages that the US public has been hearing for decades, and especially during the last decade.

The questions were:

1. Do people who live near existing nuclear facilities favor new nuclear facilities in their area? Why?
2. What electrical energy fuel sources do US residents prefer? Why?
3. What has been the impact of the Fukushima events on these preferences? Why?

Chapters 2 and 3 are essential context for the empirical studies. Chapter 2 is a summary of the story of building thousands of facilities across the USA to obtain uranium, transport it, convert it into a form useful for weapons or energy production, and then manage the accumulated waste. Chapter 3 addresses what we know about public preferences and perceptions about energy sources and waste management, especially nuclear, focusing on US surveys but also including studies from Europe. The chapter discusses the relationships between public preference and perceptions and public fears and emotions, their trust of responsible parties, their demographic characteristics, cultural and worldviews, and personal history. Chapters 4 and 5 present the empirical results. Chapter 4 summarizes the results of CRESP surveys in 2005, 2008, 2009, and 2010, that is, pre-Fukushima. Chapter 5 presents the key post-Fukushima results emphasizing the decline in public support for nuclear energy, the relationship of this change to trust and to concern about public concerns about global climate change. The final chapter addresses the implications of these results for federal government public policy, resting on the assumption that no matter what happens to nuclear power, the USA and other countries have a massive environmental legacy to manage for the foreseeable future. The chapter ends with some modest suggestions for building trust for waste management legacy programs with Fukushima as the starting point.

References

Ball M (2008) Clinton declares yucca mountain "will be off the table forever". Las Vegas Rev J. http://www.lvrj.com/news/13860977.html. Accessed February 10, 2012

Belsey J (2008) Media exposure and core values. J Mass Commun Quart 85:311–330

Blue Ribbon Commission on America's Nuclear Future (2012) Final Commission Report. January 26

Caputo A (1996) A failed experiment. The environmental forum, January/February, 17–21.

Cawley K (2010) The federal government's responsibilities and liabilities under the Nuclear Waste Policy Act. Testimony for the Committee on the Budget US House of Representatives. July 27, 2010

Clinton WJ (2011) Back to work: why we need smart government for a strong economy. Alfred A Knopf, NY

Committee on Commerce, U.S. House of representatives (2000) Incinerating cash. U.S. Washington, DC, House of representatives

CNS News (2011) president Obama defends the use of nuclear energy after Fukushima. March 16, 2011. Cnsnews.com/news/article/presdeint-Obama-defends-nculear-ernergy. Accessed February 10, 2012

Deutsch J, Forsberg C, Kadak A, Kazimi M, Moniz E, Parsons J, Yangbo D, Pierpont L (2009) Update of the MIT 2003 future of nuclear power. MIT energy initiative

Domenici P (2009) Former Senator Pete Domenic delivers speech on future of global nuclear energy. http://crespupdates.blogspot.com. Accessed December 2, 2009

General Accounting Office (2001) Fundamental reassessment needed to address major mission, structure, and accountability problems. GAO-02-51. Washington DC, GAO

Gertner J (2006) Atomic balm? The New York Times Magazine. July 16, 56–64.

Graetz M (2011) The end of energy. MIT Press, Cambridge, MA

Graetz M, Shapiro I (2005) Death by a thousand cuts: the fight over taxiing inherited wealth. Princeton University Press

Greenberg M (2009) NIMBY, CLAMP and the location of new nuclear-related facilities: U.S. National and Eleven Site-Specific Surveys. Risk Analysis, An International Journal 29:1242–1254

Greenberg M, Krueckeberg D, Kaltman M (1984) Population trends around nuclear power plants, pp 189–211. In: Pasqualetti M, Pijawka K (eds) Nuclear power: assessing and managing hazardous technology. Westview, Boulder, Colorado

Greenberg M, Krueckeberg D, Kaltman M, Metz W, Wilhelm C (1986) Local planning v. national policy: urban growth near nuclear power stations in the United States. Town Plan Rev 57:225–238

Greenberg M, Lewis D, Frisch M, Lowrie K, Mayer H (2002) The US Department of Energy's regional economic legacy: spatial dimensions of a half century of dependency. Socio-Econ Plan Sci 36:109–125

Greenberg M, Miller KT, Frisch M, Lewis D (2003) Facing an uncertain economic future: environmental management spending and rural regions surrounding the U.S. DOE's nuclear weapons facilities. Def Peace Econ 14:85–97

Greenberg M, West B, Lowrie K, Mayer H (2009) The Reporter's handbook on nuclear materials, energy, and waste management. Vanderbilt University Press, Nashville, TN

Huizenga D (2012) WM symposia 2012 and FY 2013 budget overview. Paper copy received March 15, 2012

Inhofe J, et al. (2009) Letter to secretary Steven Chu, April 29. 202009.

Jenkins-Smith H, Silva C (1998) The role of risk perception and technical information in scientific debates over nuclear waste storage. Reliab Eng Syst Saf 59:107–122

Keller B (2002) Nuclear nightmares, New York Times. May 26, 22–29, 51, 54–55, 57.

Lindell M, Earle T (1983) How close is close enough: public perceptions of the risk of industrial facilities. Risk Analysis 3:245–253

MIT (2003) The future of nuclear power. Boston, MA, MIT. http://web.mit.edu/nuclearpower. Accessed January 15, 2004

MSNBC (2010) Obama renews commitment to nuclear energy. February 16, 2000 and http://www.msnbc.msn.com/id/35421517/ns/business-oil_and_energy.

NJ DEP news (2011) Christie Administration Challenges Rule on Nuclear Waste. http://www.nj.gov/dep/docs/nrcmotion20110315.pdf. Accessed March 15, 2011

Northey H (2011) Lawmakers bitterly divided over Jaczko's future. http://www.eenews.net/public/EEDaily/2011/12/12. Accessed December 12, 2011

Nuclear Energy Institute (2002) President Bush OKs recommendation of York account mountain as disposal site for nuclear material. http://www.nei.org/newsandevents/bushokasrecomemndation. Accessed February 10, 2012

Nuclear Energy Institute (2003) Nuclear power plants vital element in President Bush's greenhouse gas reduction initiative. http://www.nei.org/newsandevents/nuclearbush/initiative. Accessed February 10, 2012

Office of Environmental Management, DOE (1995) Closing the circle on the splitting of the atom. DOE, Washington, DC

Office of Environmental Management, DOE (1996) Estimating the cold war mortgage: the 1996 baseline environmental management report, 1996, DOE/EM-0290. DOE, Washington, DC

Pegg J (2005) Bush calls for development of more nuclear power. http://www.ens-newswire.com/ens/apr2005/2005-04-28-10.asp. Accessed February 10, 2012

Pinto J (ed) (1998) The project management handbook. Jossey-Bass, San Francisco

Probst K, Lowe A (2000) Cleaning up the nuclear weapons complex: does anybody care? Center for risk management, resources for the future, Washington DC

Russell M (1997) Toward a productive divorce: separating DOE cleanups from transition assistance. Joint Institute for Energy and Environment, University of Tennessee, Knoxville, TN

Shanahan J (1995) Television viewing and adolescent authoritarianism. J Adolesc 18:271–288

Shrum L (2001) Mainstreaming, residents, and impersonal impact: testing moderators of the cultivation affects for estimates of crime risk. Hum Commun Res 27:187–215

Spangler M (1981) The role of syndrome management and the future of nuclear energy. Risk Analysis 1:179–188

Stewart R, Stewart J (2011) Fuel cycle to nowhere: U.S. law and policy on nuclear waste. Vanderbilt University Press, Nashville TN

Top-to-Bottom Review Team (2002) A review of the environmental management program.

US DOE (2006) Accelerating cleanup: Focus on 2006. Washington, DC

US Department of Energy. Department of energy releases (2001) Strategic plan (2011) May 10, 2011. http://energy.gove/articles/department-energy-releaes-2011-stratgegci-plan. Accessed may 12, 20111

Von Hippel F, ed. (2010) The uncertain future of nuclear energy. International panel of fissile materials, Research Report Number 9. http://www.fissilematerials.org Accessed October 9, 2010

Wagner R, Arthur E, Cunningham P (1999) Plutonium, nuclear power, and nuclear weapons. Issue Sci Technol. Spring, 29–33

Wallenstein P, Lynn Yang J (2011) Obama support for nuclear power faces a test. Washington Post. March 18, 2011. http://www.Washngtonpost.com/politics/obmasas-support-for-nculear-power. Accessed February 10, 2012

Wilkinson H (2006) Schmidt considers nuke waste. The Enquirer. http://news.enquirer.com/apps/pbcs.dll/article?AID=/20061029. Accessed November 1, 2006

Chapter 2
The United States Nuclear Factories

Abstract In order to arm itself with nuclear weapons, the US government mined, refined, and transported uranium products at well over 1,000 sites across the nation and in the process left levels of nuclear and chemical contamination that have become the primary mission of the Office of Environmental Management within the DOE. Nine of these sites, which are the focus of this book, have received about 90 % of the DOE's waste management budget of about $6.5 billion a year.

When the WWII ended, the USA, many European nations, Russia, and several other nations turned to atoms for peace. Over 400 operating nuclear power reactors exist in the world, with the USA, France, and Japan responsible for more than half of them. The USA has 104 operating plants at over 60 locations, and these plants not only generate electricity but also store used fuel on-site.

This chapter summarizes the decisions and actions that built the US nuclear weapons and energy factories and discusses the environmental legacy that will need to be managed into the foreseeable future at major DOE sites and at nuclear power-generating stations where commercial fuel nuclear waste is currently being stored. The costs of managing waste at several of the DOE sites already constitute the most expensive environmental management projects in the world, and the political, social, and technical challenges of the managing the nuclear factories are daunting and unprecedented.

2.1 Introduction

In 1942, the US government began a secret program to harness uranium in order to build potentially indescribably powerful weapons. In a war and fearing that their enemies might be able build and then would use these weapons, and furthermore that an invasion of Japan would cost the lives of a million American soldiers, the US government introduced the world to nuclear technology when it dropped two nuclear bombs on Japanese cities (Cooke 2009; Gosling and Fehner 1994; Groves 1962; Teller and Brown 1962).

M.R. Greenberg, *Nuclear Waste Management, Nuclear Power and Energy Choices*,
Lecture Notes in Energy 2, DOI 10.1007/978-1-4471-4231-7_2,
© Springer-Verlag London 2013

More than 70 years later, the impacts of the nuclear weapon decision are with us as strategic international, domestic political, economic, moral, occupational, environmental, and last, but not least, public perception and communication challenges. Estimates are that the US government has spent more than four trillion dollars to develop, build, test, and deliver the weapons (Schwartz 1998). For decades, mutually assured destruction (MAD) was a policy tool that came close to failing during the 1962 Cuban missile crisis (Dobbs 2008; Kennedy 1969). At the peak of the Cold War, the USA and the Soviet Union had built tens of thousands nuclear weapons and systems to deliver them (Schwartz 1998). Domestically, for decades much of public research and development dollars for energy research in the USA has gone to nuclear energy development as part of President Eisenhower's goal to use the atom for peaceful purposes (Hewlett and Holl 1989).

Would the world be any different if no one had dropped a nuclear weapon, much less built them? Managing the nuclear energy fuel cycle would have been a daunting challenge for scientists, engineers, and risk analysts. Yet, arguably, the challenge is much greater because the first use was for weapons and the first public image was the mushroom cloud. The reality of immensely destructive weapons and several failed nuclear power plants has been amplified by movies, television, and books such as *Godzilla* (1954), *On the Beach* (1959), *Dr. Strangelove* (1964), *Fail Safe* (1964), *China Syndrome* (1979), *The Day After* (1983), *The Terminator* (1984), *Atomic Train* (1999), *Atomic Twister* (2002), *Knowing* (2009), and doubtless others that I missed. Several of the movies listed above are among the most popular, having been seen by hundreds of millions of people.

An opportunity to amplify a nuclear risk image is rarely missed by creative writers and film creators. For example, the author recently read Robert Gleason's (2011) *End of Days,* which describes a successful terrorist plot to steal nuclear materials, convert them to bombs, and use the bombs to precipitate a full nuclear exchange among the world's military powers. Lastly, Chernobyl, Three Mile Island, and Fukushima have produced reality television that supports some of the ideas of the creative science fiction writers.

While public attention is drawn to nuclear disasters, the author believes that the most imposing long-range challenge is living with more subtle environmental legacies where uranium was mined, refined, enriched, fabricated, reprocessed, separated, tested, used, and otherwise incorporated into the life cycle of nuclear weapons and more recently electricity generation. In the USA and to a lesser extent other countries, thousands of factories were built to create nuclear weapons and others to turn uranium into electrical energy. An unavoidable reality left to current and future generations is that there is a legacy of nuclear materials that must be managed by stewards for more generations than I can imagine. Even if all nuclear weapons were dismantled and all nuclear power plants shut down, this legacy implies an unprecedented challenge to establish and maintain communications and cooperative relationships with those who live near sites where legacy and any new nuclear material will be stored.

This chapter provides historical, legal, and geographical context for the surveys we conducted in 2005, 2008, 2009, 2010, and 2011, primarily, although not exclusively, within 50 miles of nine DOE defense sites in the Midwest, South, and Southwest USA. This chapter briefly describes the nuclear weapons legacy and the nuclear power industry in the USA, paying far more attention to the defense sites because they set the stage for public preferences and perceptions of nuclear material as bombs and fuel.

2.2 Nuclear Weapons Factories

Most nuclear weapon components are electronic. The focus of our surveys has been on the radioactive material, not on the electronics. The awesome power of nuclear materials begins with uranium. More than 99 % of uranium atoms have an atomic weight of 238 (Greenberg et al. 2009). The remaining less than 1 % includes an isotope with a weight of 235. U-235 is more unstable than is U-238, and U-235 is a key element in nuclear weapons and power.

Building a nuclear weapon or a nuclear fuel begins with mining and milling uranium. The USA's first nuclear weapons were manufactured with uranium from Canada and Africa. During the following 40 years, about 400 uranium mines opened in the USA, almost all in Arizona, Colorado, New Mexico, Utah, and Wyoming (Fig. 2.1). It takes about a ton of uranium to produce a few pounds of uranium metal. More than 50 million tons of ore were mined for the war effort, leaving a massive volume of waste. The mined uranium ore had to be crushed, leached with acids, and then milled into so-called yellowcake, which has a much higher concentration of uranium oxide then the unrefined ore. These processes produced residuals commonly called mill tailings (they look like sand) that contain radioactive by-products, as well as metals. These initial steps in bomb-making generate over 95 % of the waste volume. The radioactive waste material has not been in a nuclear reactor and hence is not as hot as nuclear material that has been, yet this waste can migrate and expose people and ecosystems. In 1978, the US Congress recognized that these initial by-products of the bomb-making process had not been adequately secured and required them to be more thoroughly stabilized, remediated, and managed.

Yellow cake was shipped to refineries that concentrated the milled uranium into products that could be used to produce nuclear weapons. Fernald (OH), (Fig. 2.2), one of the sites we surveyed, was a major refining site. Fernald's products were transported to nearby Paducah (KY) (Fig. 2.3), Portsmouth (OH) (Fig. 2.4), and Oak Ridge (TN) (Fig. 2.5) (included in our surveys), where the refined material was further separated and concentrated to increase the proportion of U-235 to 20 % or even more for weapons, naval reactors, and other uses. These facilities primarily used centrifuges to concentrate the more fissionable elements. The volume of waste was much less than produced by mining and milling, and the residuals were more radioactive. The processes required extensive use of metals and solvents, leading to localized contamination of nuclear and other chemicals at the sites.

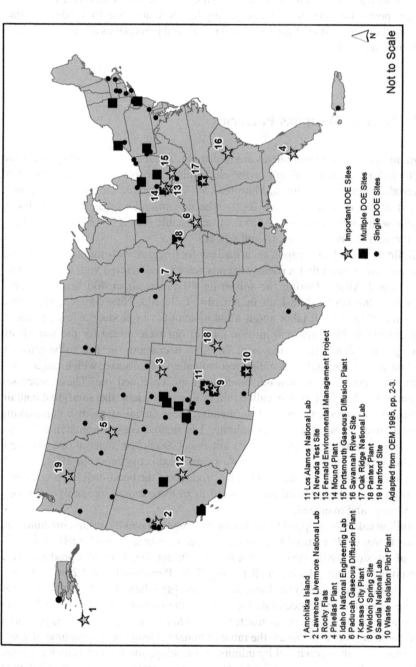

Fig. 2.1 DOE defense sites

1 Amchitka Island
2 Lawrence Livermore National Lab
3 Rocky Flats
4 Pinellas Plant
5 Idaho National Engineering Lab
6 Paducah Gaseous Diffusion Plant
7 Kansas City Plant
8 Weldon Spring Site
9 Sandia National Lab
10 Waste Isolation Pilot Plant

11 Los Alamos National Lab
12 Nevada Test Site
13 Femald Environmental Management Project
14 Mound Plant
15 Portsmouth Gaseous Diffusion Plant
16 Savannah River Site
17 Oak Ridge National Lab
18 Pantex Plant
19 Hanford Site

Adapted from OEM 1995, pp. 2-3.

☆ Important DOE Sites
■ Multiple DOE Sites
● Single DOE Sites

Not to Scale

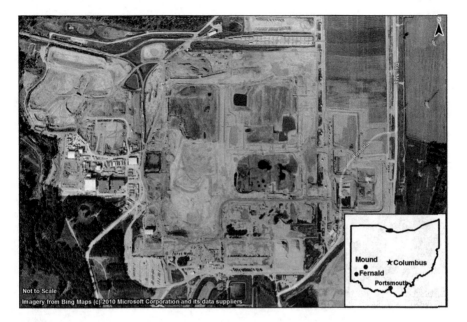

Fig. 2.2 Fernald site

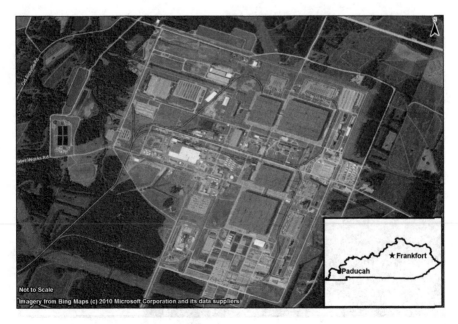

Fig. 2.3 Paducah site

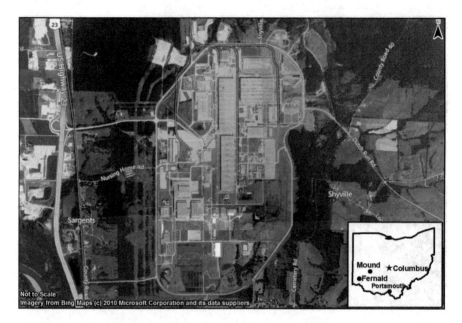

Fig. 2.4 Portsmouth site

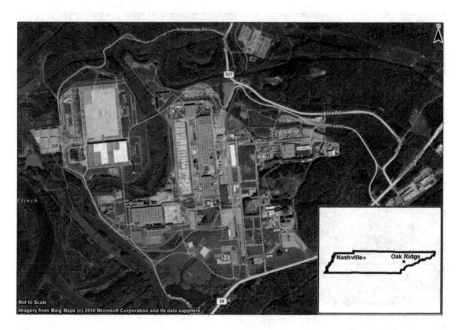

Fig. 2.5 Portion of Oak Ridge site

Fig. 2.6 (a) Portion of Hanford site. (b) Hanford site regional context

Further refining converted the nuclear elements into weapons grade material. For example, at Fernald, uranium hexafluoride gas was converted into crystals and then into a uranium metal. This metal became reactor fuel or plutonium targets at Hanford (Fig. 2.6a, b), Oak Ridge (Fig. 2.5), Rocky Flats (Fig. 2.7), and Savannah River (Fig. 2.8) (all included in our surveys). All of these facilities have had issues

Fig. 2.7   Rocky Flats site

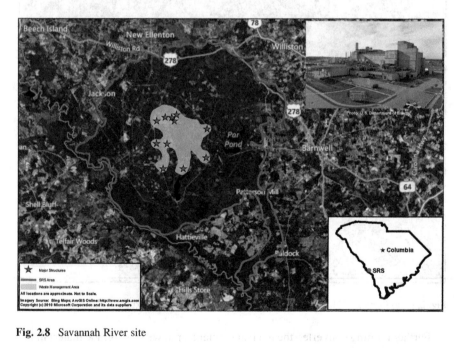

Fig. 2.8   Savannah River site

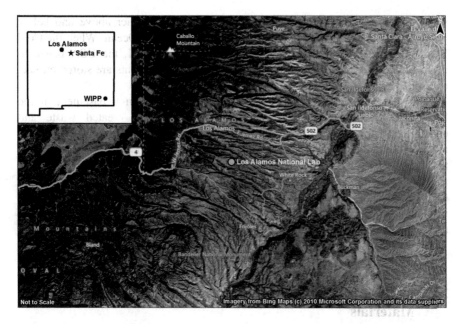

Fig. 2.9 Los Alamos regional context

with airborne dust, leakage from waste sites, and Fernald and Rocky Flats were decontaminated and raised when they were closed.

Plutonium U-239 is a human-made form of uranium that merits special attention. Created in nuclear reactors by bombarding U-238 with neutrons from the disintegration of U-235, it is then separated from other radioactive fission elements by introducing acids. The waste from this process has proven to be a major costly environmental challenge for half a century at the Hanford and Savannah River sites and to a lesser extent Idaho. High-level waste from 14 reactors was stored in massive tanks that were not designed to withstand corrosion for half a century. The two most expensive waste management projects in the world are the vitrification (glassification) factories at Savannah River and Hanford. The first at Savannah River was completed at a cost of $2.5 billion in 1996. The second at Hanford is still being built with a budget of $12.3 billion and that budget is increasing (Dininny 2011).

Much of the plutonium from the reprocessing plants was shipped to the Rocky Flats site (located about 15 miles from downtown Denver in the foothills of the Rocky Mountains) where it was manufactured into warhead components, so-called plutonium triggers or pits. The management of plutonium at Rocky Flats required the utmost care, which at this and other sites was not always a highest priority.

Nuclear weapons were primarily designed at the Los Alamos (NM) (Fig. 2.9), Livermore (CA), and Sandia (NM, near Albuquerque) national laboratories. Assembly of the weapons was at the Pantex facility near Amarillo (TX). Most weapons tests took place at the Nevada Test Site, but other above and below grounds tests were in Alaska, and on islands in the Pacific Ocean. When the Cold War ended with the breakup of the Soviet Union in 1991, the USA began to disassemble the weapons at Pantex. Some weapon components are stored on-site, and some are shipped to Savannah River and Oak Ridge.

Summarizing, beginning with the Manhattan project in a basement at the University of Chicago, building nuclear weapons has created waste and contaminated thousands of sites across the USA and some outside the USA. By today's standards, occupational and environmental management practices during this almost 40 year period were not acceptable. The most obvious evidence is soil and water contamination. Management of this legacy has not been as compelling to the US public as Three Mile Island, Chernobyl, Fukushima, and international intrigue about nuclear weapon proliferation, but it is important to those who live near the legacy sites.

2.3   Federal Government Management of Defense Nuclear Materials

During World War II, the US military and contractors managed the nuclear weapon life cycle. After the war, the federal government turned to civil management. In 1946, the Atomic Energy Commission (AEC) began to administer the growing number of nuclear weapons and study the feasibility of nuclear power and human and ecological health impacts of nuclear materials. In 1975, facing political pressure as promoter and regulator of atomic power, the AEC was replaced by two federal agencies. The Energy Research and Development Administration (ERDA) became the promoter and the Nuclear Regulatory Commission (NRC) the government manager of the private nuclear power industry. The DOE replaced ERDA in 1977 (Fehner and Holl 1994; Office of Environmental Management 1995).

2.3.1   Office of Environmental Management

The newly created DOE faced an unprecedented challenge of stabilizing the most dangerous wastes in massive underground storage tanks. In 1989, the Office of Environmental Management was established with the primary mission of managing so-called high-level waste (Fehner and Holl 1994; Office of Environmental Management, DOE 1994, 1995, 1996a, b, 1997). Legally, high-level waste is radioactive material produced by reprocessing spent fuel from reactors and irradiated targets. This material includes highly radioactive elements that decay rapidly and others that will be hazardous for thousands of years. Most of this high-level defense legacy waste came from producing plutonium, and as noted above, almost all of it

Fig. 2.10 WIPP site

has been stored in massive underground tanks at Hanford and Savannah River. The Defense Waste Processing Facility was built at Savannah River, which mixes the radioactive waste pumped from the tanks with a molten glass mixture in large stainless steel canisters. At Hanford, which has even more high-level waste, vitrification facilities have been under construction.

DOE-EM is also responsible for transuranic waste, which is waste with higher atomic numbers than uranium, such as americium and plutonium. Some of these elements have decay periods of thousands of years. Transuranic waste shipments to the Waste Isolation Pilot Plant near Carlsbad, New Mexico, began in 1989 (Fig. 2.10).

While these sites have been the primary endeavor the Office of Environmental Management (OEM), the EM group has three other tasks. It is responsible for waste management at other sites, as well as waste from reactor research, associated science, and special wastes from events such as the Three Mile Island accident in March 1979. When sites are officially closed, they must be remediated, which includes landscaping and the installation of monitoring systems, drainage, and other systems that will permit long-term stewardship, and surplus facilities must be managed and materials in the DOE's inventory secured and safeguarded. EM is also deeply engaged in international efforts to reduce the chances of nuclear proliferation. For example, DOE has a central role in managing the down blending of nuclear materials purchased from decommissioning the inventory of nuclear weapons that were built by the former Soviet Union and by the USA (see Chap. 1

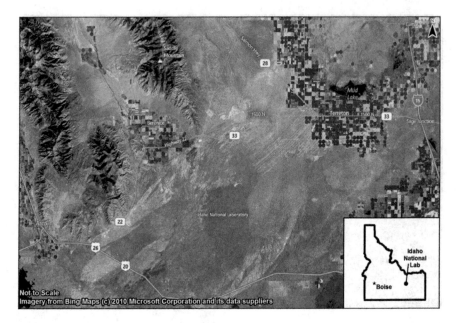

Fig. 2.11 Idaho National Laboratory site

discussion of proliferation). Lastly, DOE has responsibility for low-level, mixed radioactive and chemical wastes and the uranium mill tailings.

Nevertheless, the primary challenge is the high-level wastes, which means a great deal of effort directed at Hanford, Savannah River, and also Oak Ridge, Fernald, Idaho (Fig. 2.11), Rocky Flats, Los Alamos, and WIPP. Since 1989 about 90 % of EM's annual budget of about $6.5 has been spent at these eight locations, about half at Hanford and Savannah River (US Department of Energy 2012. See below for more details about the sites and chapter for an update and look forward on expenditures).

2.3.2 Legal Framework for Defense Wastes

The vast majority of DOEs nuclear factories are closed or will be. Only 17 out of more than 130 are still active. Some will remain and continue to have primarily a national security mission along with other science, technology, and waste management ones. The DOE's programs are regulated and guided by multiple federal laws and rules (see Stewart and Stewart 2011; US Environmental Protection Agency 2012, for presentations of the laws). I briefly summarize these.

In 1970, the National Environmental Policy Act required all federal agencies to prepare an environmental impact statement (EIS) of proposed actions that could impact the environment. The idea was to require the agency to consider a variety of options before making a final choice, including a no-action one. Every major DOE action at its sites requires an EIS or environmental assessment (actions that are assumed not to substantially threaten the environment) (Greenberg 2012).

The Comprehensive Environmental Response, Compensation and Liability Act (CERCLA), also known as Superfund, was aimed at locating, containing, and cleaning up thousands of abandoned landfills, factories, and other properties that threaten public health and the environment. Passed in 1980 and reauthorized and amended in 1986, Superfund was primarily aimed at chemical contamination, but almost every major DOE site has been a Superfund site, for example, Fernald, Idaho, Hanford, Mound, Oak Ridge, Pantex, and Savannah River.

The Resource Conservation and Recovery Act (RCRA) was passed in 1976 and amended in 1984 and 1988. RCRA regulates the operations of existing private and government facilities, including DOE's, that generate, store, and dispose hazardous waste. RCRA also mandates that sources, such as the DOE, attempt to reduce the volume of waste and hazardousness of the waste by applying source reduction, recycling, and material substitution.

While these three are the primary laws, the DOE's waste management programs are also guided by other federal and state laws. I briefly summarize the federal ones in alphabetical order. The Atomic Energy Act of 1946 (amended in 1954 and later) requires that DOE manages all nuclear materials at its facilities and that the Nuclear Regulatory Commission (NRC) manages those generated by private organizations. The Clean Air Act of 1963, as amended, requires that every emitter obtain permits that limit emissions and/or requires the use of specific technologies and fuels.

The Clean Water Act of 1972 controls the emission of pollutants into navigable water bodies. Emitters must obtain a permit for direct discharge or reach an agreement to discharge their waterborne waste through a permitted public waste treatment facility. The Nuclear Waste Policy Act (NWPA) of 1982 and amendments require the DOE to design, locate, and build a geologic repository for the long-term disposal of defense high-level radioactive waste and spent nuclear fuel from commercial nuclear reactors. This mandate has not been accomplished, and many residents and key elected officials of Nevada have been adamantly against the Yucca Mountain facility (see sections of Chaps. 1, 5, and 6).

The Occupational Safety and Health Act (OSHA) of 1970 requires employers like the DOE and the contractors to identify risks to workers and eliminate or minimize their exposure. The Safe Drinking Water Act (SDWA) of 1974 attempts to protect potable drinking water sources. As a provider of drinking water at some sites, DOE is responsible for complying with SDWA. The Toxic Substances Control Act (TSCA) of 1976 requires testing for the presence of specific chemicals and restricting public exposures to these. Last on this short list of laws is the Uranium Mill Tailings Radiation Control Act of 1978 that directs the DOE to stabilize and control uranium mill tailings. For the DOE this implies cleaning up dozens of sites and adjacent properties. Also, please note that the DOE is also responsible for complying with many other federal statutes about protecting archaeological and cultural resources, endangered species, transportation of dangerous materials, low-level radioactive waste, oil pollution, noise, and land use.

Part of the DOE's challenge is to deal with legal definitions of nuclear waste (Greenberg et al. 2009; Stewart and Stewart 2011). We know that radiation comes from water, rocks, soil, as well as from outer space and that there are three types of ionizing radiation. Alpha particles are relatively large and slow moving and are stopped by skin and paper. Smaller and fast beta particles pass through skin paper but can be stopped by a relatively thin layer of metals such as aluminum. Gamma radiation travels at the speed of light, and a substantial shield of lead, concrete steel, or other material is required to stop it.

Some of these radioactive materials decay to nonhazardous levels in a period that is consistent with demonstrated human capacity to manage infrastructure. Others will be hazardous in tens of thousands of years. Each form of radioactive imposes toxicity and length of management challenges.

For DOE nuclear waste management is guided by federal regulations not necessarily directly linked to risk. For example, high-level waste is radioactive material from reprocessing of used nuclear fuel, including the fuel itself, liquid waste, and solid waste derived from the liquid. This waste must be handled by remote control. Transuranic wastes are elements that are heavier than uranium, emit alpha radiation, and are produced during weapons fabrication chemical processing and the assembly of reactor fuels. Low-level radioactive waste is defined as any radioactive waste not classified as high level, transuranic, or uranium mill tailings. While low-level radioactive waste typically is clothing, dirt construction rubble, protective clothing, tools, and so forth, some low-level radioactive waste can be risky, and yet environmental management of it does not necessarily reflect the risk.

2.4 Geography of Nuclear Defense Sites

In a complex of over 5,000 facilities (Office of Environmental Management 1994, 1995, 1997), the DOE has focused its programs on about 130 (Fig. 2.1) located in almost every state but clustered in New Mexico and California, Colorado, Ohio, New Jersey, New York, and Pennsylvania. Measured by budget, DOE has focused on less than 20 (Fig. 2.1, larger circles). The list includes the three national labs at which the weapons were developed: Lawrence Livermore National Laboratory (CA), Los Alamos (NM), and Sandia (NM). Also on the list is the Nevada Test Site (Fig. 2.12). Savannah River (Fig. 2.8), Hanford (Fig. 2.6a, b), and Idaho (Fig. 2.11) had multiple defense missions and the high-level radioactive waste. Rocky Flats, where plutonium triggers were built, Mound (Fig. 2.13), Paducah, Pinellas, Portsmouth, Burlington, Kansas City, and Weldon Spring and Fernald where various components were manufactured and refined are part of the group of less than 20 high priority sites. The last two are Pantex, where weapons were built and now disassembled, and WIPP, where transuranic waste is managed.

The key places that influenced our selection for sampling were where used (spent) fuel is located. Hanford, SRS, Idaho, and West Valley had over 99 % of spent fuel. The relatively small amount of high-level waste at West Valley has been

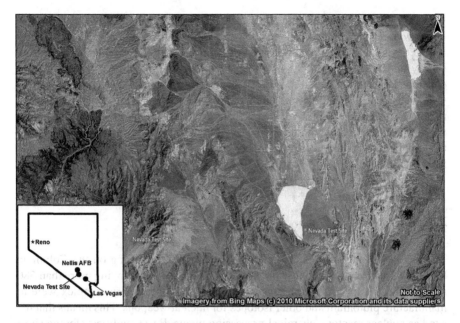

Fig. 2.12 Nevada test site regional context

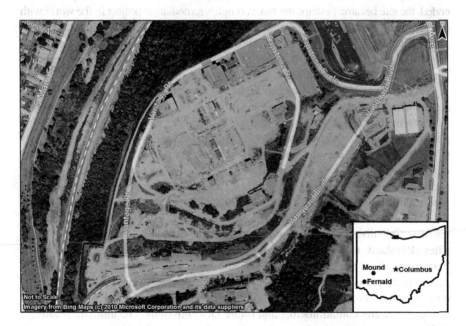

Fig. 2.13 Mound site

vitrified (glassified), and it no longer represents a threat compared to the much larger volumes at Savannah River, Idaho, and especially Hanford.

Oak Ridge represents a complex set of facilities with remediation activities and energy and nuclear research activities. Los Alamos represents a small densely developed site in an arid area that has been a major contributor to DOE nuclear weapons science. WIPP was not included in our first surveys, but has become a major waste management site. Rocky Flats and Fernald were included in several surveys because both were in the process of closing, and we wanted to understand the populations' preferences and risk beliefs compared to the DOE sites that would have ongoing activities.

The following summarizes each of the sites in our surveys.

2.4.1   Hanford

The DOE Hanford site lies in southeast Washington, bordering on the Columbia River, not far from the border with Oregon (Fig. 2.6a, b). Occupying more than 580 square miles, its primary initial tasks in the DOE bomb factory complex were to manufacture plutonium and other isotopes for nuclear weapons. This means that the site has nuclear reactors, chemical processing plants, R&D facilities, storage sites, and many nuclear and chemical waste management facilities. When production ended, the site became perhaps the most complex remediation project in the world with decontamination, deactivation, stabilization, remediation, and decommissioning.

The site also has several other research missions, including the Pacific Northwest National Laboratory, multiple other research laboratories, and natural resource preserves operated by the federal government and the State of Washington.

A key driver in site management is the so-called Hanford Federal Facility Agreement and Consent Order, better known as the Tri-Party Agreement signed in 1989 by the DOE, EPA, and the Washington Department of Ecology. The agreement commits the DOE to working with these parties and local governments to plan future use of the site. The Hanford Citizens Advisory Board, a group with a remarkably diverse set of participants from both Washington and Oregon, represents local residents, site employees, local government organizations, not for profits, and Native American Tribes. The Hanford plan has future use objectives for the site as a whole and areas of the site and offers a variety of variations that reflect the incredible diversity of the populations. Notably, the area includes several small cities (Richland, Kennewick, Pasco) and the Umatilla and Nez Pez Indian tribes. Exchanges about the future use of the site have been contentious with distinct perspectives voiced by the Native Americans, the local residents, and the DOE over remediation of the high-level waste in over 100 underground tanks, remediation of soil, groundwater contamination, and future DOE missions. The ultimate challenge at Hanford, however, is the nuclear waste legacy stored in tanks constructed between 1943 and 1964, some of which have been closed and replaced by other sturdier tanks, which in turn are to be closed as the vitrification facility is completed and activated.

2.4.2 Idaho National Laboratory

The 890-square-mile Idaho site is located in southeast Idaho about 30 miles west of Idaho Falls (Figs. 2.11). The site is best described as nine manufacturing and waste management clusters set down amid semiarid high desert terrain. The site initially was a gunnery range for the US Navy. Remnants of structures where guns from battleships were tested are still visible, and one can imagine massive main battleship rounds being fired from them. In 1950, the Atomic Energy Commission took over the site, and both developed and tested nuclear reactors and related technology. From the 1970s into the 1990s, the Idaho site added waste minimization and management, environmental restoration, engineering, and various nonnuclear-related activities related to national security and energy efficiency, biotechnology, and other technologies. At one time or another during its 60 years history, the site has had a large radioactive waste management cluster, a breeder reactor program, and several test areas. Among the prominent residuals at the site is granulated high-level waste stored in underground stainless steel bins, as well as transuranic, low-level, and mixed wastes.

During the 1990s, the site designed a future land use plan that included participation from DOE staff, the site citizen advisory board, the state and the federal EPA, the Bureau of Land management and the US Forest Service and Park Service, members of the local business community, and Shoshone-Bannock tribes. The essence of the plan is that the federal government will continue to own and manage the site, and that DOE and non-DOE research and industrial technology activities are encouraged around the existing cluster of activities. These activities will be related to nuclear technology, waste management, national security, and other engineering and science and technology programs focused on the environment. In the short term, however, the site has major remediation, decontamination, and decommissioning programs.

2.4.3 Savannah River

DOE's 310-square-mile Savannah River facility lies entirely in South Carolina. Its western border is the Savannah River and the Georgia lies directly across the river (Fig. 2.8). The Vogtle nuclear power plant is across the river in Georgia.

SRS is a visual experience, which the author characterized as Bombs and Butterflies (Bebbington 1990; Greenberg et al. 1997). Ten percent of the site has massive facilities associated with the production of plutonium and tritium gas for nuclear weapons. The site had five nuclear reactors, 51 massive underground storage tanks to contain high-level waste, and assorted other large industrial facilities and ponds for waste management. It has the world's largest vitrification (glassification) facility that mixes high-level waste from the tanks with molten glass and pours the aggregate into large stainless steel containers that weigh 4,000 pounds. The facility is expected to produce 6,000 canisters by the time it closes in 2019 (Office of Environmental Management 2009).

SRS also has a new mixed oxide fuel facility built in conjunction with AREVA that converts weapons grade material (as required by a bilateral agreement between the USA and Russia) to mixed oxide fuel destined to be burned in nuclear power plants. At the time this chapter was being written, several power plants were being considered. MOX fuel shipped from SRS would require special security precautions and would doubtless draw attention from local media and elected officials at Savannah River, along the transportation route, and at the final destination.

It is fair to characterize the Savannah River site as epitomizing large complex industrial manufacturing and waste management facilities. It also is fair to characterize it as a beautiful gem of forests that have not been disturbed for well over half a century and are now managed by the Forest Service for timber production. The site also has ample fish and wildlife and is a major and ecological research laboratory.

Given this amazing variety of existing land uses and the reality that the site will be open for future national security energy missions, the Savannah River site has faced a challenging exercise of developing and adjusting a Savannah River future use plan that can accommodate a breadth of land uses.

2.4.4 Oak Ridge

The Oak Ridge Reservation is the most complex DOE site because of its multiple missions. Its budget historically has been divided among environmental, energy, and weapons-related functions, and it includes three DOE installations (Fig. 2.5). The K-25 core with over 300 buildings sits on about eight square miles, focusing on environmental management, including design, development and demonstration, and education programs. The Oak Ridge National Laboratory sits on about 38 square miles in the middle of the site with basic science functions around nuclear fission, fusion, energy conservation, and a variety of other research activities associated with environmental management. The eastern approximately seven square miles of the site is occupied by the Y-12 core area, which lies adjacent to the city of Oak Ridge. Historically this area focused on dismantling weapons and storing materials including uranium, as well as environmental management and decontamination. In addition to these three core areas, a great deal of the site is undeveloped and essentially serves as a nature preserve, which was codified in 1980 as a National Environmental Research Park.

Befitting its complex and varied missions in the DOE factory complex, Oak Ridge has many radioactive and hazardous contamination issues ranging from enriched uranium to asbestos that will require DOE funding for many years. The site will also continue its varied nuclear and energy functions.

2.4.5   Los Alamos

Located on a visually striking plateau with mesas, canyons, and steep slopes, the desert that encompasses the 43-square-mile Los Alamos National Laboratory (LANL) site is sparsely developed (Fig. 2.9). The developed parts include laboratories, science facilities, burial trenches, and buildings primarily used in the development of nuclear weapons, and these are scattered around the site with a large buffer areas between them. LANL contains about two dozen disposal areas that have the residuals of radioactive and hazardous chemicals including plutonium, uranium, tritium, as well as metals and pesticides and a long list of other chemicals used in scientific research. The site has identified over 2,000 contaminated locations, including septic tanks, pits, trenches, and surface impoundments. Of particular concern are so-called firing sites where explosive charges were detonated resulting in dispersion of natural and depleted uranium, beryllium, and other heavy metals. Given its remote location (approximately 60 miles from Albuquerque), its ongoing nuclear weapons missions spread across the site, and the lack of water resources on the site, I cannot imagine LANL not remaining an important federal government asset for the foreseeable future.

2.4.6   Waste Isolation Pilot Plant

Waste isolation pilot plant (WIPP), located near Carlsbad, New Mexico, stores transuranic wastes 655 m (2,150 ft) below ground in a salt basin with a 250-million-year history (Fig. 2.10). WIPP's initial shipment came from Los Alamos on March 26, 1999, and the plan is to receive waste for 25–35 years on trucks and trains. The DOE chose this site after a location in Lyons, Kansas, failed and New Mexico elected officials suggested exploration of the Carlsbad area.

The site is 41.4 km^2 (16 square miles), and the DOE has invested about $20 billion in the site and has been strongly supported by New Mexico elected officials. Legislation mandates that the site concentrate on transuranic wastes from DOE's defense sites, not accept high-level radioactive waste, and the current site requirements have a maximum waste load. However, in light of the tenuous position of Yucca Mountain, there surely will be efforts to broaden WIPP's mission, perhaps by adding another repository adjacent to the current one. Jenkins-Smith et al. (2011) observed that the WIPP site is increasingly supported and suggests that the DOE's efforts to gain public acceptance are working with some residents.

2.4.7   Closed Sites: Fernald, Mound, and Rocky Flats

In 1951, the Atomic Energy Commission began manufacturing high-grade uranium on a 1,050-acre rural site that is partly in Hamilton County and partly in Butler County, Ohio, about 20 miles northwest of Cincinnati. When the manufacturing ended at the Fernald site (Fig. 2.2), the DOE began the cleanup that eventually cost $4.4 billion (Vartabedian 2009). The small site now is covered with trees, grasses, and wetlands and commemorated by a holographic postcard that shows the site when it was active and now in its final ecological state.

Downtown Denver clearly is visible from the DOE Rocky Flat site located in the foothills of the Rocky Mountains (Fig. 2.7). From 1952 to 1989, it produced nuclear weapon components from plutonium and other elements, and it conducted research associated with its nuclear weapons mission. After a safety review that disclosed unacceptable practices at the site, plutonium operations stopped in 1989 and resumed, only to end operations in 1992. During its almost 50-year history, Rocky Flats attracted a great deal of public and political attention and was a high priority for cleanup and closure.

In the author's experience at various DOE sites, the Rocky Flats citizen's advisory board was among the most active and influential. The small size of the site (approximately 10 square miles) and suburban location meant that local townships, chambers of commerce, citizen groups, and local businesses landowners had personal states and were deeply involved in future use planning. The site has become an ecological preserve.

The DOE's Mound facility located in Miamisburg, Ohio (Fig. 2.13), is intended for reindustrialization. The 306-acre facility is located approximately 10 miles southwest of Dayton and produced detonators and igniters for nuclear weapons. DOE found over 400 contaminated locations with plutonium, tritium, and thorium, as well as various industrial cleaning agents. The site had landfills, overflow ponds, and several instances of off-site releases, including radioactive material, and immediately adjacent to it is a city golf course, residential and active cultural areas, as well as industrial uses. Given the urban location of the site, considerable effort has been made in cleaning it up, and the City of Miamisburg's goal is to turn it into industrial and other commercial activity.

2.4.8   Economic Dependence and Public Preferences

Chapter 3 will show that public preferences and perceptions are influenced not only by fears but also by economic benefits. In the context of these major DOE sites, Table 2.1 compares total DOE expenditures in six of these regions during the mid-1990s. These estimates are averages, varied by year, and are not current. With these caveats noted, the numbers were chosen to show remarkable economic dependence on DOE funding when the DOE was allocating tens of billions of dollars for

Table 2.1   DOE contributions to the regional economies of six major regions during 1994–1996

DOE site	All DOE investments as % of regional GRP	EM investments as % of regional GRP	EM as %
Rocky Flats	5.2	2.6	51
Los Alamos	26.2	4.4	17
INEL	20.0	16.9	84
Savannah River	16.5	8.2	49
Hanford	16.5	14.1	85
Oak Ridge	14.4	3.9	27

environmental cleanup. With an average annual budget of about $6 billion during this period in 1992, 20 % went to Hanford, another 20 % to Savannah River, 13 % to Oak Ridge, 7 % to Idaho, and another 7 % to Rocky Flats, leaving 33 % for the remainder of the USA (see Chap. 6 for more recent allocations).

The denominator for the calculations in Table 2.1 was the gross regional product of the economic region calculated by the author and colleagues from federal government data, and the numerator was the total DOE budget allocation during those years. The DOE investment in the Rocky Flats site during that time is illustrative of what we would expect to find in a densely developed metropolitan region and would be similar for the Sandia, Brookhaven, and other DOE sites in metropolitan regions, that is, low dependency on DOE. In strong contrast, the direct DOE investment is a major portion of the gross regional product of the Los Alamos, Idaho, Hanford, Savannah River, and Oak Ridge areas. These data suggest that there should be significant local public preference for DOE investments in these regions and the continuation of major environmental management programs. Chapters 4 and 5 will show this preference at some but not all of the DOE sites.

2.5   Nuclear Power

From the late 1930s through the successful development of the hydrogen bomb, the USA and USSR competed to develop and build the capacity to deliver nuclear weapons. With the US President Eisenhower's 1953 Atoms for Peace speech as a prominent signal, the momentum gradually shifted to technology that would use the immense heat released by fission to create abundant and inexpensive electrical energy. In 1951, a fission reactor at what is now the INEL site (reactor now housed in a small museum at INEL) produced a small amount of power for Arco, Idaho.

The initial major push for nuclear power was for nuclear submarines. Under US Admiral Hyman Rickover, the pressured water reactor (PWR) was developed to power nuclear submarines. The Nautilus, launched in 1954, was able to go on extended missions with its PWR. In 1957, a 60-MWe nuclear power plant was built at Shippingport (PA) that operated until 1992. The first moderately sized plant (250 MWe), built by Westinghouse, was Yankee Rowe (MA). General Electric built the first boiling water reactor (BWR), which started operating in 1960. A decade later, utilities were ordering PWRs and BWRs of 1,000+ MWe, mostly the former.

The UK, Canada, and France pursued their own paths to nuclear power, and using its nuclear weapons technology base, the Soviet Union developed a set of prototypes and then multiple RBMK reactors, including the one at Chernobyl. Not until the 1990s were a third generation of reactors produced. In 1996, the Kashiwazaki-Kariwa reactor, located on the west coast of Japan on the Sea of Japan, became the first advanced boiling water reactor. Kariwa has experienced earthquakes with some resulting damage and was closed for 21 months.

2.5.1 Nuclear Power Life Cycle

Recognizing that life cycle of every nuclear power plant is unique, nevertheless, this section very briefly summarizes the process of converting uranium into electrical energy and managing the residuals. Please note that this synopsis focuses on the nuclear fuel life cycle. It should be noted that a life cycle view is consistent with public risk perceptions and beliefs, which are not limited to nuclear reactor accidents. The public's perceptions of nuclear energy include mining, transportation, waste management, as well as the management of the power plants (Greenberg and Truelove 2011).

The nuclear power plant life cycle starts with mining uranium or recovery of uranium, which is blended to a level required for a nuclear fission in the reactor. The manufactured fuel is produced in small thumb-sized pellets and then placed in zirconium cladding and then into larger fuel assemblies (Greenberg et al. 2009). The assemblies are joined in a reactor core. The reactor core is radioactive and is surrounded by a protective shield. With some notable exceptions, the reactors are enclosed in large concrete domes to prevent damage to the reactor by outside events and to protect the outside environment from problems with reactor operations. Nuclear power plants also have major and minor plumbing capacity to prevent the reactors from overheating, which can lead to a melting of the heated fuel as in the cases of Chernobyl, TMI, and Fukushima. Operators control the reaction with valves, pumps, and various other systems, and the operators in the USA are now required to train on mock-ups of the actual reactor they operate so that they can practice and be prepared for scenarios ranging from small to very serious.

Plant operators initiate controlled fission in the reactor core. The enormous heat created is transferred to a coolant and creates steam that activates massive turbines that convert kinetic energy produced by the turbine into electrical energy. At preplanned times, some assemblies are removed because they are no longer efficient. These are removed to large pools of water where they are left until their heat decreases. Eventually, the so-called used fuel is placed in concrete casks and stored at each site. The geologic repository within Yucca Mountain (Fig. 2.14) is partly

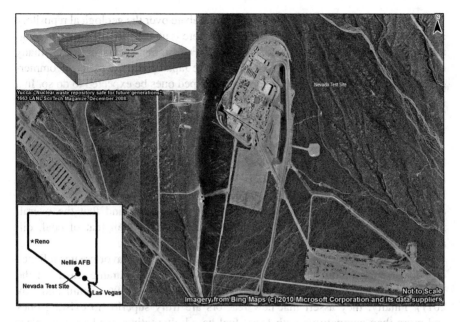

Fig. 2.14 Yucca Mountain site

constructed, but as of this writing is not in use nor planned to be in use in the near term future, if at all.

2.5.2 Nuclear Power Debates

Beginning in the late 1960s and escalating during the mid-1970s, the world's industrial and industrializing nations have engaged in an unprecedented debate about the wisdom of nuclear power as a base electricity-generating source (Brook and Lowe 2010; Cravens 2007; Falk 1982; Gertner 2006; Massachusetts Institute of Technology 2003, 2011; Nuclear Energy Institute 2012; Walker 2006). Opponents characterize nuclear power as a threat to people and the environment. While they focus on nuclear power reactor accidents at Three Mile Island, Chernobyl, and Japan, they include evidence that begins with uranium mining and includes processing, refining, transport, long-term storage and disposal, possible theft of nuclear materials, terrorist threats, and dependence of nuclear power plants on large and continuous water supplies. New types of reactors are characterized as adding to complexity and risk. Opponents characterize the economics of nuclear power as extraordinarily expensive compared to alternatives to design, construct, and protect and that nuclear power has only reached its current status because of massive government subsidies that should have been allocated among other energy sources and/or conservation. Furthermore, detractors assert that the cost of decommissioning is high and is likely to increase. Indeed, opponents use the ongoing debate over the geological repository at Yucca Mountain in Nevada to illustrate that there is no long-term solution to long-term management of wastes. After the 2011 Fukushima events in Japan, many opponents argued that if an advanced society like Japan could not manage commercial nuclear reactors, then how could less developed ones be expected to do so. In a post-Fukushima article, the influential barometer, the *Economist* (Nuclear Power 2011), stated that the nuclear looks dangerous, unpopular, expensive, risky, and is replaceable.

Proponents of nuclear energy characterize it as a sustainable energy source that will reduce carbon emissions. For some countries, nuclear power plants mean less dependence on international suppliers who can cut off coal and gas shipments. Proponents assert that nuclear fuel is relatively inexpensive, wastes are manageable compared to those from coal and natural gas production, and that the overall reliability and safety records of nuclear energy is better than that of coal, oil, and gas.

While the US Yucca faculty is not operational, those who favor greater reliance on nuclear energy point to the opening and storage of transuranic wastes at the WIPP facility (Heaton 2010; National Research Council 1996; Timm and Fox 2011). Finally, they assert that new reactors are truly superior to existing ones and more than competitive with fossil-fuel-based alternatives for base load electricity needs.

Proponents assert that the Nuclear Regulatory Commission requires risk assessments for every plant, compulsory worker training and testing, makes genuine efforts to determine if the plants can respond to terrorist attacks, and many other requirements that have no equivalent at coal, oil and natural gas plants, and mines. They point to accidents at coal mines, the Gulf Coast platform blowout, tanker groundings, a some cases of accidents at liquefied natural gas plants, the threat posed by LNG tankers, pipeline breaks, and many more to counter the assertion that no nuclear plants should be built, that existing ones should not be renewed, and that the technology is too dangerous.

2.5.3   Uncertain Status and Future of Nuclear Power

Whatever the merit of the decision to build nuclear power plants, less than a year after the Fukushima events in 2011, the world had 433 nuclear reactors in 31 nations, which is 11 less than the peak of 444 in 2002. These represent 367 GW (a gigawatt is 1,000 MW), (International Atomic Energy Agency 2006, 2012).

The USA, France, and Japan have more than half of the world's nuclear energy capacity, and in total, nuclear power plants account for about 15 % of the world electricity. France is the most dependent country, with 75 % of its electrical energy from nuclear reactors. This compares with 20 % in the USA. The US proportion is exceeded in Armenia, Belgium, Bulgaria, Czech Republic, Finland, Germany, Hungary, Japan, South Korea, Slovakia, Sweden, Switzerland, and the Ukraine.

While France, Japan, and the USA have clearly invested in nuclear power, and other less populated and less populated countries have staked a great deal on nuclear power, still other industrial nations have just as clearly not. Austria, Ireland, and Italy, among others, have no active nuclear power facilities, and in the wake of the Fukushima events, Germany and Switzerland have decided to close their plants by 2022 in the first case and by 2034 in the second. Japan's nuclear power policies are not to rely on the sources, but they have changed their plans, as can those who are currently promoting nuclear energy.

Yet, the most important decisions about nuclear power will be made in China, India, and other rapidly developing nations. The author visited China in August 2010 and spoke with senior government officials and members of the Academy of Sciences who were clear about choosing nuclear power rather than coal. China then with 13 nuclear power plants, others under construction, talked about adding 50 more. Among the 150 planned new nuclear power plants, almost two-thirds were to be in China, India, Japan, and Russia (International Atomic Energy 2006, 2012).

Notably, following the Fukushima event, the International Energy Agency reduced its estimate of additional nuclear generating capacity to be built by the year 2035 in half. Yet, I do not believe that anyone knows what will happen, including nations that currently are antinuclear power. The USA, for example, has relicensed many of its existing reactors for up to 60 years, but will this continue in light of Fukushima, and what happened in Japan which had made a major commitment to nuclear energy? Will Germany follow-through with the announcement to close all it nuclear plants? Realistically, we do not know if by the year 2030 nuclear energy will reach close to 700 GW or drop to levels below 2012.

Public preferences and perceptions will surely have some impact on nuclear power decisions. Chapter 3 shows that public preference for nuclear power expansion is closely associated with their government position. France and Germany have markedly different positions. Are government officials who read public opinion polls taking political positions about nuclear power that the plurality of constituents supports? Or is the public taking cues from its national government's position? Likely, a feedback loop is working in both directions.

2.5.4 Geography of Nuclear Power, Other Electrical Energy Sources, and DOE Nuclear Facilities in the United States

Figure 2.15 shows the locations of the 104 operating nuclear power plants in the USA (US Nuclear Regulatory Commission 2011). The mountain and plains states have very few. Over 40 % are in Illinois, Pennsylvania, South Carolina, New York, Alabama, and Florida (each has at least five). Given the reality that there is no permanent geological repository nor an interim one, the used fuel from the plants is stored in large pools or in concrete casks at the sites and will remain there for the near term. Used fuel, which was not known to be a major concern or even known to

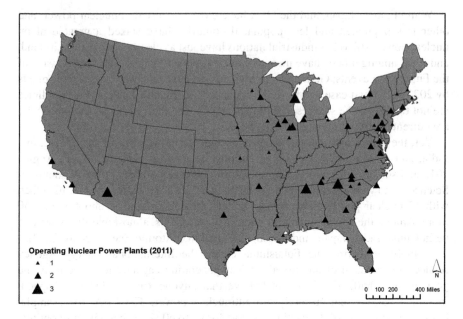

Fig. 2.15 Operating nuclear power plants in the USA, 2011

many people prior to Fukushima, is now of concern to people, and hence, residents should be concerned not only about reactor and transportation problems but also about used fuel (Macfarlane and Ewing 2006, see Chap. 5).

In 2005, I asserted at a meeting that national polls about preferences for nuclear waste management and nuclear power were potentially misleading for characterizing the views of those who live with the risks and benefits of sites that produce the energy and manage the residuals. For example, DOE's largest major waste management sites were chosen to avoid populated areas. After considering urban sites such as a nuclear plant across the river from the United Nations building in New York City and another close to the City of Philadelphia, the Nuclear Regulatory Commission guidelines required nuclear power plants to be at least 20 miles away for an urban center (Greenberg et al. 1986). That decision to separate people from urban areas has been functionally undermined by population migration, for example, the Indian Point (NY) plant is now an urban site.

The cases of coal and natural gas are quite different. Natural gas production and liquefied natural gas plants are clustered in a few regions, but natural gas is widely distributed through pipelines to homes and centralized facilities, and therefore, natural gas should be familiar to people (Energy Information Administration 2010). Coal was shipped to urban load centers and some people will remember coal burning to heat individual homes. However, in parts of the USA, many sites formerly used for burning coal have been shut down or burn gas, and hence, the geography of large coal plants should be somewhat different from the population distribution.

Table 2.2 Correlations of population distribution by state in 2010 with electrical energy sources and DOE waste sites, 2010

Attribute	Spearman rank correlation	States with most
Population	–	CA,TX,NY,FL,IL
Natural gas consumption	0.828**	TX,CA,LA,NY,FL
Nuclear power plants	0.753**	IL,PA,SC,NY,FLA,AL,NC
Coal plants	0.616**	OH,IN,MI,IL,PA
DOE waste management sites	0.335	CA,NM,OH,CO,NY
DOE major waste sites	0.244	OH,NM,MO,13 had 1
Gas production	0.014	TX,WY,OK,LA,CO

**Correlation statistically significant at $p < 0.01$, *$p < 0.05$

As context for our site-specific surveys focused around nine DOE sites, and Table 2.2 shows the correlation between the distribution of people, electrical energy sources, and DOE sites by state in the USA.

Table 2.2 shows the expected high correlations with natural gas consumption and significant correlations with nuclear power plants and coal plants. California, Texas, New York, and Florida have many people and a great deal of gas consumption. The five most populated states have between 4 and 11 nuclear power plants. Coal plants are disproportionately in Midwestern states that also have substantial populations. Some DOE facilities are also found in populated states, but the larger sites are not in the most populated states. Gas production, as distinct from consumption, is geographically clustered in a north to south band from Wyoming to Louisiana. In other words, the results of a national random sample about energy preferences and risk perceptions would not represent the views of people who live near major DOE nuclear waste management facilities. Chapter 3 summarizes what we know about public preferences and perceptions in USA.

References

Bebbington W (1990) History of DuPont at the Savannah River plant. El DuPont DeNemours and Company, Wilmington, DE

Brook B, Lowe I (2010) Why vs. why: nuclear power. Pantera, Sydney, Australia

Cooke S (2009) In mortal hands: a cautionary history of the nuclear age. Black, Inc., Collingwood, Australia

Cravens G (2007) Power to save the world: the truth about nuclear energy. Knopf, NY

Dininny S (2011) DOE review raises concern about Hanford plant. http://www.seattlepi.com/news/article/DOE. Accessed November 22, 2011.

Dobbs M (2008) One minute to midnight: Kennedy, Khrushchev, and Castro on the brink of nuclear war. Knopf, NY

Energy Information Administration (2010). Natural gas annual, 2009. Released December 28, 2010. Natural Gas production, Transmission, and Consumption, by State, 2009. Table 2, page 4. http://www.eia/gov. Accessed December 26, 2011

Falk J (1982) Global fission: the battle over nuclear fission. Oxford University Press, NY

Fehner T, Holl J (1994) Department of Energy, 1977–1994, a summary history. DOE/HR-0098. USDOE, Washington, DC

Gertner J (2006) The nuclear option. The New York Times Magazine. July 16, 36–47, 56, 62, and 64

Gleason R (2011) End of days. Forge Books, NY

Gosling F, Fehner T (1994) The Manhattan project: making the atomic bomb. US Government Printing Office, Washington, DC

Greenberg M (2012) The environmental impact statement after two generations: managing environmental power. Routledge, New York

Greenberg M, Truelove H (2011) Energy choices and perceived risks: is it just global warming and fear of a nuclear power plant accident? Risk Anal 31:819–831

Greenberg M, Krueckeberg D, Kaltman M, Metz W, Wilhelm C (1986) Local planning v. national policy: urban growth near nuclear power stations in the United States. Town Plann Rev 57:225–238

Greenberg M, Lowrie K, Krueckeberg D, Mayer H, Simon D (1997) Bombs and butterflies: a case study of the challenges of post cold-war environmental planning and management for the United States' nuclear weapons sites. J Environ Plann Manage 40:739–750

Greenberg M, West B, Lowrie K, Mayer H (2009) The reporter's handbook on nuclear materials, energy, and waste management. Vanderbilt University Press, Nashville, TN

Groves L (1962) Now it can be told: the story of the Manhattan project. Da Capo Press, New York

Heaton J (2010) The Carlsbad/WIPP history of transuranic disposal in salt. Presentation to the Blue Ribbon Commission on America's Nuclear Future Disposal Subcommittee. July 7, 2010. http://www.energyca.org/pdf/CarlsbadHistory.pdf. Accessed December 17, 2011

Hewlett R, Holl J (1989) Atoms For peace and war, 1953–1961: Eisenhower and the Atomic Energy Commission. University of California Press, Berkeley, CA

International Atomic Energy Agency (2006) Nuclear power reactors in the world. http://www.iae.org.org/mtcd/publications/pdf. Accessed January 9, 2012

International Atomic Energy Agency (2012) Nuclear power global status. 49(2), http://www.iae.org/Publications/Magazines/bulletin/Bull492. Accessed January 9, 2012

Jenkins-Smith H, Silva C, Nowlin M, deLozier G (2011) Reversing nuclear opposition, evolving public acceptance of a permanent nuclear waste disposal facility. Risk Anal 31:629–644

Kennedy R (1969) Thirteen days: a memoir of the Cuban missile crisis. WW Norton, NY

Macfarlane M, Ewing R eds. (2006) Uncertainty underground, Yucca Mountain and the nation's high-level nuclear waste. Massachusetts Institute of technology press, Cambridge (M Café)

Massachusetts Institute of Technology (2003) The future of nuclear power. MIT Press, Cambridge, MA

Massachusetts Institute of Technology (2011) The future of the nuclear fuel cycle. MIT Press, Cambridge (MA)

National Research Council (1996) The waste isolation pilot plant, a potential solution for the disposal of transuranic waste. National Academy Press, Washington, DC

Nuclear Energy Institute (2012) U.S. nuclear power plants. http://www.nei.org. Accessed January 7, 2012

Nuclear Power: When the steam clears. The economist. March 24, 2011. http://www.economist.com/node/18441163. Accessed January 9, 2012

Office of Environmental Management, DOE (1994) Environmental fact sheets. USDOE, Washington, DC

Office of Environmental Management, DOE (1995) Closing the circle on the splitting of the atom: the environmental legacy of nuclear weapons production and what the department is doing about it. DOE/EM-0266. DOE, Washington, DC

Office of Environmental Management, DOE (1996a) Estimating the cold war mortgage: the 1996 baseline environmental management report, 1996, DOE/EM-0290. DOE, Washington, DC

Office of Environmental Management, DOE (1996b) Charting the course, the future use report, DOE/EM/0283. DOE, Washington, DC

Office of Environmental Management, DOE (1997). Linking Legacies. http://www.em.doe.gov/publications/linklegacy.aspx. DOE/EM-0319, DOE, Washington, DC

Office of Environmental Management, DOE (2009) Defense waste processing facility news. http://www.srs.gov/general/news/factsheets/processingfacilitydwpf.pdf. Accessed January 7, 2012

Schwartz S (ed) (1998) Atomic audit: the costs and consequences of U.S. nuclear weapons since 1940. Washington, D.C., US Nuclear Weapons Cost Study Project. http://www.brookings/edupress/books/1998/atmoc.aspx. Accessed January 7, 2012

Stewart R, Stewart J (2011) Fuel cycle to nowhere. Vanderbilt University Press, Nashville (TN)

Teller E, Brown R (1962) The legacy of Hiroshima. Doubleday, Garden City, NY

Timm C, Fox J (2011) Could WIPP replace Yucca Mountain? Nuclear Engineering International. http://www.neimagazine.com/story.asp. Accessed December17, 2011

US Department of Energy (2012). Budget and performance. http://www.cfo.doe.gov/budget. Accessed January 7, 2012

US Environmental Protection Agency (2012) Laws and regulations. http://www.epa.gov/lawsregs. Accessed January 7, 2012

US Nuclear Regulatory Commission (2011) Operating nuclear power reactors. As of May 19, 2011. http://www.nrc.cog/info-finder/reactor. Accessed December 29, 2011

Vartabedian R (2009) Toxic legacy of the cold war. Los Angeles Times. Articles.latimes.com/2009/oct/20/nation/na-radiation-fernald. Accessed January 7, 2012

Walker JS (2006) Three Mile Island: a nuclear crisis in historical perspective. Berkeley, CA

Chapter 3
Public Stakeholders: What We Know and Expect

Abstract Hundreds of surveys have sampled public preferences for different energy sources and associated perceptions. Far fewer polls have focused on waste management. The foundation for in-depth investigations rests on five sets of respondent attributes: (1) concern about the options, marked especially by strong feelings and emotions; (2) trust of responsible parties; (3) demographic characteristics, including race/ethnicity, age, and socioeconomic status; (4) culture and worldviews; and (5) personal history, including familiarity with sites and benefits associated with living near sites. The better studies show variation over time and place in support for nuclear and other energy sources and waste management policies. European surveys, for instance, show fascinating differences among adjacent countries that appear to parallel national energy policies. Initial reports show that the Fukushima nuclear-related events have caused a decrease in public support for nuclear energy that is also associated with decreasing trust of those responsible for managing waste sites.

3.1 Introduction

Hundreds of surveys have asked questions about nuclear-related issues, and yet we are far away from an adequate understanding of public preferences and perceptions about nuclear technology and even further away on how to interpret them, especially after the events at the Fukushima nuclear facilities in Japan in March 2011. The vast majority of surveys have four shortcomings. Almost all are about nuclear power, and rarely have there been questions about the entire nuclear fuel cycle from mining to waste management. We have data on how many people want new nuclear power plants and greater reliance on nuclear energy in multiple countries, but we do not know much about public views regarding obtaining the uranium, managing the wastes, and transporting nuclear materials. A second limitation is that the overwhelming majority of the surveys are part of much larger polls that include questions about the latest political issues, sporting events, energy prices, and other "newsworthy" issues; in other words, they lack context. Sometimes three or four questions are needed to understand what people perceive, and only one question leaves uncertainty. Third, nuclear-related questions in the vast majority of surveys are not built around theoretically important questions (trust, affect, economic benefit) that would help us better understand the relevance of the findings, and in many cases, demographic data have not been collected that allow us to measure variations by age, income, and other standard indicators. The last limitation is, with rare exception, the surveys are national samples, rather than samples of places where the benefits and risks associated with nuclear technology disproportionately will be directly felt. In a democracy, theoretically everyone is equal, but this is not the case when it comes to the benefits and risks of alternative energy options because the risks and perhaps benefits are clustered and as shown in Chap. 2 are only moderately to weakly associated with the distribution of the US national population.

This chapter is divided into three parts. It begins by characterizing the body of surveys on nuclear technology. Next, it reviews key theory that should help us understand public preferences and perceptions of nuclear technology, and third, it reviews some of the most informative surveys that have been conducted. It is important that readers realize that we have been selective, focusing on the USA and the European Union.

3.2 Surveys, 2005–May 2011

We searched "nuclear energy public opinion polls" and similar combinations for the period 2005–2011. A total of 148 surveys were found. Table 3.1 categorizes them by year, source, and focus. One hundred were in 2005, 2006, and 2008. About 60 % were standard media polls, and most of the remaining were by well-known polling organizations, notably the Eurobarometer for the European Union, Gallup, Pew, and Harris, and the Nuclear Energy Institute.

The vast majority were about gas prices and contained one or two nuclear power-related questions in the surveys, and nothing about nuclear waste management. Nuclear power was also a subject in surveys about energy alternatives. Only about 25 % of the surveys were primarily about nuclear energy, and few had even a single question about nuclear waste. The typical survey was a random digit dialing landline phone survey of adults with about 1,000 respondents, although some were face-to-face surveys. The USA population as a whole was almost always the sample target.

I would characterize the vast majority as atheoretical, that is, have no grounding in risk-related theories, nor were the surveys built around a series of formal research questions that are answered through hypothesis-driven designs and analyses. Accordingly, the remainder of this chapter summarizes the body of theory that has guided the research reported in this book and then provides greater detail on a set of studies that merit attention because they have all or at least some of the

Table 3.1 Sources of public opinion polls about nuclear-related issues, 2005–May 2011

Source	Attribute	Number of surveys
Year	2011 (through May)	18
	2010	11
	2009	10
	2008	36
	2007	9
	2006	34
	2005	30
Sponsor	Media (ABC, AP, CBS, CNN, FOX, LA Times, NBC, USA Today)[a]	85
	Research & polling organization (Eurobarometer, Gallup, Pew, Harris)	37
	Nuclear Energy Institute	9
	University	6
Major focus of questions	Energy choices	36
	Gas prices	64
	Global climate change	2
	Nuclear power	37
	Oil drilling	6
	Radioactive waste	3

[a]ABC news, Associated Press sometimes with a partner, CBS news, CNN, Fox News, Los Angeles Times sometimes with a partner, NBC news, USA Today sometimes with a partner

following characteristics: long time series, post-Fukushima results, high-quality design, and focus on nuclear waste management. As context for Fukushima, the chapter closes with a review of the public opinion literature that reported on the Chernobyl and TMI incidents. Many of these studies are high quality.

3.3 Theory

Over 3 decades ago, it became clear that people's preferences did not follow actuarial expectations. Actual and estimates of death and injuries were only somewhat associated with public perceptions.

3.3.1 Affect and Worry

The "psychometric paradigm" was the most important initial theoretical proposal to address the lack of association between what people were supposed to be concerned about and what they actually were worried about. Psychologists demonstrated that public preferences and risk beliefs were driven by feelings, emotions, experiences, and present knowledge. Each person constructs a mental model of risk that includes such factors as catastrophic potential, controllability, dread, equity, impacts on future generations, uncertainty, and other factors. Researchers gathered data about this set of factors and using multivariate statistical methods grouped them into two or three factors. The most common factors are "dread' and "unknown" (Slovic 1987; Cha 2000; Bronfman and Cifuentes 2003; Xie et al. 2003). Dozens of risks have been classified, including nuclear war, nuclear power, and nuclear waste, as well as auto accidents and global warming. Hazards that are high in dread and tend to be unknown are at the top of the public's high-risk list. Smoking tobacco and traffic accidents kill and injure tens of thousands of people a year in the USA. However automobile accidents and smoking are common and considered less dreadful. In the USA, nuclear power is typically the most dreaded and unknown risk, or among the most dreaded and unknown, despite the fact that experts can show that driving intoxicated or with a cell phone and not using a seat belt are much more likely to lead to a death or serious injury than operating a nuclear waste management or nuclear power plant.

In a pioneering study, Slovic (1987) found that out of 30 different risks, nuclear power ranked first among college students and members of the League of Women Voters. Cha (2000) studied a similar set of risks in South Korea, finding that among 70 risks, nuclear weapons/war, nuclear weapon tests, and nuclear reactor accidents ranked 1, 2, and 4, respectively. In addition, radioactive waste management, transportation of nuclear materials, and nuclear power plants ranked 11, 12, and 19, respectively. Bronfman and Cifuentes (2003) observed that nuclear weapons had the highest dread among 54 risks in a sample of Chile's residents, and nuclear power ranked fourth. In China, Xie et al. (2003) observed that respondents ranked nuclear war number 1. Yet, nuclear power ranked 27 out of 28 risks. These are but a few of the many international analyses that demonstrate why many people worry more about nuclear technology than morbidity and mortality data would suggest.

A common thread in this literature is the affect heuristic, asserting that feelings must be activated before people connect with an issue, and, once activated, positive and negative feelings influence individual decisions and even may lead people to choices that are contrary to their personal best interest. The affect heuristic has been used by scholars to try to better understand people's concerns about pesticides; reactions to biotechnology; choices to purchase insurance, continue smoking, make charitable contributions, and select dictionaries; and to help guide other preferences and behaviors (Alhakami and Slovic 1994; Dickert and Slovic 2009; Finucane et al. 2000a,b; Hsee 1996; Hsee and Kunreuther 2000; Salvadori et al. 2004; Slovic 2001; Slovic et al. 2004, 2007; Small and Loewenstein 2003; Vastfjall et al. 2008).

The Yucca Mountain controversy is interpretable through the psychometric paradigm and the affect heuristic. Kunreuther et al. (1990) asked 1,001 residents of Nevada if they would be willing to accept the permanent nuclear waste repository, which was located about 90 miles north of Las Vegas. The surveys proposed offering annual tax rebates of $1000, $3000, or even $5000 per year for 20 years. Relatively few respondents indicated that they would change their opposition to the facility, even with the promise of tax relief. The controversial Yucca Mountain facilities illustrate the powerful role of feelings, emotions, and images. Not only was there the image of nuclear risk, but also having the reputation as the nuclear waste capital of the USA has been a major concern among Nevada residents who prefer the Las Vegas gambling and entertainment image. Also, many Nevadans consider the choice of Yucca Mountain as unfair/inequitable. That is, the original US government legislation promised a national search and the selection of multiple sites. The search process was cut short, and Yucca Mountain was chosen. Accordingly, Nevadans have good reasons to have strong negative emotions about the site, irrespective of the science-related issues (Macfarlane and Ewing 2006).

While there are good reasons to have negative feelings and worries about nuclear technology, there are reasons to have positive ones, especially among people who live near some nuclear sites, such as those included in our surveys. Public risk beliefs, perceptions, and preferences change. Recent negative events strongly influence public perception (Koren and Klein 1991; Siegrist and Cvetkovich 2001; Skowronski and Carlston 1989), which is one reason that the Fukushima events are so important to study. The further away in time we get from the Three Mile Island and Chernobyl events, the less negative emotions and dread should exist. However, as discussed below, some data contradict that expectation.

What is particularly notable about our surveys is the sampling areas disproportionately are near former weapons facilities. While negative feelings dominate in most applications of the affect heuristic, the author did not necessarily expect this to be the case at locations with a long history in the nuclear defense complex. Presumably, some people would have negative feelings and emotions, such as fear, hate, disgust, fury, anger, distrust, and worry. Others should have pleasant feelings, such as good, happy, glad, trust, safe, and pleased. What we did not know was how many would have positive and how many negative emotions and feelings. We suspected that there would be at least a majority on the positive side because of the history of these regions. For these regions, described in Chap. 2, these nuclear facilities have been a source of jobs, economic stability, and pride in their national defense mission.

While the mushroom cloud is a powerful image (Weart 1992), I did not think many people would have these images in the six areas decades after the end of World War II and the Cold War. Some who are familiar with the sites might see the vast open spaces, forests, mountains in several cases, and large majestic landscapes that dominate these sites as beautiful and desirable (Greenberg et al. 1997). We also expected images of massive cooling towers, reactor buildings, and other structures. Having visited these sites, the author also expected images of people working, especially scientists in lab coats, and working with robots.

While the psychometric paradigm and affect have been the core theoretical foundations for risk preference and perception research, studies show that that "dread" and "unknown" factors explain about 20 % of the variation in risk perception (Sjöberg 2003).

3.3.2 Trust

Trust is the logical second choice as a correlate of preferences and risk perceptions. If people believe that responsible parties are competent, fair, acting in good faith, and also share their values, they are more likely to be more supportive of nuclear-related initiates than their counterparts who are not trusting (Earle 2010; Earle and Cvetkovich 1995; Nye et al., 1997; Pew 1998; Poortinga and Pidgeon 2003; Siegrist and Cvetkovich 2000; Cvetkovich and Roth 2000).

The role of trust is hard to isolate because there are interactions between trust and other factors. For example, Sjoberg (2002) found that there is only a relatively weak association between perception and trust because respondents feel that scientists do not know all the impacts of technology. Siegrist et al. (2000) observed that people who were knowledgeable about risks did not rely on trust to assess a hazard. Those who were unfamiliar with the hazard relied heavily on their trust of the operators and mangers of the hazard.

The impacts of trust, knowledge, morality, and benefits on public preference for biotechnology were jointly examined by Knight (2007). Knowledge was an indirect predictor, notably mediated by trust, which in turn was a weaker predictor than morality and benefits. Vandermoere (2008) considered perception of the need to clean up soil contamination, observing that beliefs about how much experts knew were predictive.

Earle's (2010) review paper about trust distinguishes between "relational" trust, which focuses on intentions and values, and "confidence," which is about experience, ability, and competence. Siegrist (2010) adds that researchers do not have an unambiguous definition of trust and how it should be measured. Both of these leading experts emphasize the need to pose clear questions.

3.3.3 Demographic Attributes

Measurable demographic characteristics have repeatedly been observed to be associated with risk-related preferences and perceptions (Finucane et al. 2000b,c; Flynn et al. 1994; Greenberg 2005; Stern 2000). A "white male effect" has been observed; in other words, affluent, educated, politically influential white males are more supportive than relatively poor, less formally educated, and less politically connected African Americans and Native Americans and women. Flynn, Slovic, and Metz (1994) examined a large set of environmental risks with a national US sample, finding that white men perceived most risks to be less worrisome and much more acceptable than their nonwhite male and female counterparts. Interesting variations on the demographic correlates have appeared in recent years. For example, Rivers et al. (2010) looked for and did not find a similar effect in the African American community. Greenberg and Schneider (1995) found greater female concern about industrial and environmental risks in neighborhoods classified by the respondents as excellent or good. However, in neighborhoods classified as fair or poor, the male–female difference was not apparent. Olofsson and Rashid (2011) looked for a male white effect in Sweden, a nation without marked gender-based differences in socioeconomic status, education, and political power. The authors did not find a white male effect but did find an immigrant–nonimmigrant effect.

3.3.4 Cultural Worldviews

Broad views and values (Douglas 1970; Kahan et al. 2007) have been suggested as a possible underlying foundation of the white male effect and other demographic differences (Peters and Slovic 1996). Two dimensions are proposed. Those who are hierarchical believe that power and resources should be allocated according to social status. They contrast with those who are egalitarian and believe in a more equal allocation of resources (Gross and Rayner 1985; Rayner, 1992). Second, those who are individualistic feel that society should be organized according to competitive outcomes and emphasis should be on independent action. They contrast with communitarians who believe in group membership and social interdependencies (Rayner 1992). Those who are hierarchical and individualistic have been found to be more supportive of nuclear power than their counterparts (Greenberg and Truelove 2010; Peters and Slovic 1996). In the US context, hierarchical and individualist overlaps affluent male whites.

3.3.5 Personal History

Experience influences perceptions and preferences. Many attributes fit under personal history, beginning with parental influences that not only shape feelings about specific technologies but impart a sense of optimism or pessimism to many people and a feeling about the importance of the local environment. Personal history also includes work history and geographical attributes. For example those who work for a nuclear technology company, are well paid, are familiar with the site, reap benefits from retail sales in their jurisdictions from site workers, and in other ways benefit from nearby nuclear facilities should be more favorably disposed to a site than their counterparts who are not familiar with the site and rely on media reports for information (Greenberg 2005, 2009a,b; Greenberg et al. 2012).

3.4 More Detailed Presentations

3.4.1 European Studies

3.4.1.1 Eurobarometer Surveys

The European Commission for the European Union frequently surveys residents of member states about EU-wide issues. These EU-wide surveys have four important attributes. The first is a sample size of over 25,000. The EU asks people in more than two dozen nations the same questions with the same sampling protocol. The Eurobarometer surveys typically involve 1,000 samples in each of the larger member states and 500 each in Cyprus and Malta. This sample size allows comparisons with greater confidence than in nearly any other survey with a typical sample size of only 600–1,200. A second strength is that some of the Eurobarometer surveys have important contextual questions, such as how important energy issues are compared to crime, war, jobs, and other national issues. This allows the results to be set in proper policy and urgency contexts. A third advantage of the EU surveys is that they publish a full report, typically around 150 pages that includes key findings, useful graphics, and detailed appendices, including the questions. The fourth advantage is that the surveys focus on a key policy issue, such as energy policy from generation to storage rather than merely a single aspect, such as nuclear energy. Frankly, the author believes that North America would benefit from a similar process applied to the NAFTA nations.

Here four EU surveys in 2005, 2006, 2008, and 2010 are highlighted. The 2005 survey was entitled "Radioactive Waste" (European Commission 2005, 2006, 2008, 2010), but it is much broader insofar as it includes questions about nuclear energy, public knowledge, and other important issues. I highlight several findings of the 2005 survey. One-fourth of respondents said that they were well informed about nuclear waste issues, with a range of 51 % in Sweden to 15 % in Portugal. Respondents were asked questions that the surveyors used to score actual knowledge. The average of correct answers was 53 %, which frankly was better than this author had anticipated (Greenberg and Truelove 2010).

With regard to nuclear energy and waste, 37 % were in favor of nuclear energy, 55 % were against, and 8 % had no opinion. In 2005, international differences were striking, with a high of 65 % in Hungary to 8 % in Austria.

The full set of Eurobarometer surveys (European Commission 2005, 2006, 2008, 2010) report six policy-relevant findings regarding nuclear technology. One is that citizens are not well informed about nuclear energy or waste. Their preferences and perceptions are more associated with their national government's policy than with their personal attributes. Those in favor of nuclear energy were disproportionately men, more formally educated, employed as mangers, self-declared politically to the right, and knowledgeable about nuclear science facts. This finding is entirely consistent with the white male affect in the USA described above and below.

A second interesting finding is that nuclear waste perceptions and preferences are connected to nuclear power. Almost 4 in 10 respondents said that they would be more positively disposed to nuclear power if there was a permanent and safe plan to manage nuclear wastes. Disproportionately these "changeable" respondents were highly educated young respondents. In contrast, many other respondents saw no solution to the waste management issue. Related to this second finding was that over 90 % of respondents felt it was urgent to find a permanent solution to the radioactive waste management issue.

A third notable observation was that EU residents were aware of some benefits of nuclear energy, such as reducing greenhouse gas emissions, diversifying fuel use, and reducing dependence on international imports of fuels. The last finding is the surveys report a gradual increase of support for nuclear power in the EU during the study period so that by 2010, about equal proportions were for and against nuclear power. A similar pre-Fukushima increase was also observed in the USA (see Chaps. 4 and 5).

3.4.1.2 University-Based Surveys in Europe

A group of British professors, primarily at Cardiff University, published a series of papers focusing on the psychological underpinnings of preferences and perceptions of nuclear power. Using a year 2010 national survey of 1,822, Corner et al. (2011) found almost the exact same level of support (35 % in 2010 and 36 % in 2005). The key observation was that UK residents were reluctant to support nuclear energy, but they would conditionally support it if renewable energy, conservation, and other approaches do not reduce greenhouse gas emissions (see also Pidgeon et al. 2008; Parkhill et al. 2010). Another paper by this group (Parkhill et al. 2011) investigated the possibility that humor is allowing people to express their fears and criticize authority about the nuclear power and waste issues without confrontation. Both of these papers underscore the role of emotions and affect.

Visschers and Siegrist (2012) investigated how the Fukushima events changed acceptance of nuclear power and related these changes to perceived benefits and risks associated with nuclear power and trust in German-speaking areas of Switzerland. A key objective of this work was to determine how public perceptions of benefits and risks were changed by the event and how these were associated with change in acceptance. The authors assumed that perceived benefits would influence perceived risks because the former include measurable quantities such as reduction of greenhouse gas emissions, whereas the risks are much less easy to quantify.

The design was to send a paper-and-pencil survey instrument to a randomly chosen group of German-speaking households in Switzerland in September 2010. After follow-up, the authors received 1,233 responses. The same individuals were asked to respond to a shorter version of the questionnaire two weeks after the March 2011 Fukushima events. A total of 929 surveys were returned. Some of these were not used because different respondents completed the second survey and in some cases there was too much missing data. The analysis was done on 790 responses to both surveys.

Using a variety of multivariate statistical tools, the authors report that both acceptances of nuclear plants decreased and that benefits and trust decreased significantly. On a 7-point scale, in 2010, the average values to three acceptance questions ranged between 4.05 and 4.59. After the event, these dropped to between 3.12 and 4.11, an average of 20 %. Not surprisingly, average values of perceived risks increased and average values of perceived benefits dropped. The authors conclude that the event confirmed people's risk perceptions but notably changed the benefit side more than the risk side. In other words, respondents viewed nuclear plants as much less beneficial than they had in the past.

A 2009 study by Keller, Visschers, and Siegrist (2012) explored images and emotions about nuclear power in Switzerland. The authors report that those opposed to nuclear power had more specific and a greater diversity of negative images than those who were in favor of nuclear power, and they conclude the complexity of negative images will be more difficult to overcome. Notably, Chap. 4 will show that when the sample is of people who live near DOE's major nuclear facilities, then positive images are more frequent than negative ones (see also Greenberg 2012).

3.4.2   United States Surveys

3.4.2.1   Nuclear Energy Institute–Bisconti

The most persistent effort to monitor public opinion has been by the Nuclear Energy Institute, which has hired Bisconti Associates (2007, 2008, 2009, 2010a, b) to study residents who live near the sites, almost always within 10 miles. The series began in 1983 and most recent survey was post-Fukushima. Bisconti Research with NOP (formerly RoperASW) uses a random digit dialing sampling approach and asks approximately 1,000 US adults age 18 and older a set of questions about nuclear power plants and energy. Some of the questions have been asked every year, and others are of more recent origin. Until the Fukushima event, they report a clear trend in the data since 1996 toward more public support for nuclear energy.

The surveyors ask people if they strongly favor, somewhat favor, somewhat oppose, or strongly oppose the use of nuclear energy as one of the ways to provide electricity in the USA. In 1983, 49 % were in favor of nuclear power. This increased to 52 % in 1985 and ranged between 49 and 55 % until 1994. Given that the margin of error for the survey is plus or minus three percentage points, I cannot conclude that this change represents a trend. But even before Fukushima, there are some interesting and surprising results. Crude oil prices dropped by more than 50 % between 1981 and 1986. The Chernobyl reactor meltdown occurred in April 1986. The first Gulf War occurred in 1991 and was marked by a slight increase in oil prices. If public opinion about nuclear energy in the USA responded to other events, we should have seen some movement in these data. That is, the public should have become more interested nuclear when fossil fuel prices increased. But there was no evidence of that in the Bisconti reports.

According to these NEI-sponsored surveys, the striking change in public preferences happened between 1995 and 1998. Forty-six percent of the population favored using nuclear fuel to produce electricity in 1995. One year later, it was 49 %, and the proportion jumped to 61 % in 1998, and then to 62 % in 1999. There was a slight increase in the price of oil between 1998 and 1999, but nothing like the change in public preferences. Beginning with the new millennium, the proportion favoring nuclear energy had steadily increased, to more than 70 % in the May 2005 survey, and then to 74 % in April 2010. These increases correspond to the terrorist attacks in 2001 and a substantial increase in fuel prices. But the reasons for the large increase between 1995 and 1998 are not obvious to this author.

The Bisconti data show support for siting nuclear plants at existing locations, the so-called choose locations at major plants (CLAMP) (Greenberg 2009a). In 2005, 69 % said it was acceptable to build a new plant near an existing site, an increase from 57 % in 2003. About 64 % said there was a nuclear power plant in the state where they lived; hence, these data show that a strong not in my backyard (NIMBY) response did not exist among these populations.

In the period June 11–18, 2011, Bisconti Research (2011) conducted a post-Fukushima survey of 1,152 (18 samples near each of 64 nuclear power plants) that compared 2011 answers to many of the same questions asked in 2009 and earlier years. Without going over every question, the result is that the public is slightly, but not markedly more uncomfortable with nuclear energy after Fukushima than before it. For example, in 2009, 92 % said the nuclear energy was important in meeting the nation's electricity needs, and in 2011, the number was 87 %. Other notable observations were that in 2011, 86 % said that we should renew the license of nuclear power plants that continue to meet federal safety standards, which was down from 93 % in 2009. In 2009, 86 % said we should keep the option to build more nuclear power plants in the future; the proportion fell to 79 % in 2011 after Fukushima. When asked in 2011 about their "general impression" of the nuclear power plant nearest them, 86 % said favorable compared to 90 % in 2009.

Bisconti asked what people associate nuclear energy with. In 2009, about 70 % said efficiency, reliability, and clean air, and in 2011, these numbers were about 62 %. In contrast, in 2009, 35 % said nuclear energy was a solution for climate change, and in 2011, the proportion was 29 %. On a scale of 1–7, where 1 represents very unsafe and 7 represents very safe, 88 % of respondents in 2009 rated their nearby plant 5–7, and this proportion fell to 83 % in 2011. Lastly, there were similar slight decreases in confidence in the companies to operate a nuclear power plant safety, the company's ability to protect the local environment, and slight decreases in the job and economic benefits of the plant to the community. Across all of the proportions expressing confidence in the company were well over 85 % in 2009 and fell 2–3 % to about 82 % in 2011.

With the caveat that we have not been able to obtain the raw data, that the population lives within 10 miles of an existing nuclear power plant, and that these results do not address those that live near DOE waste management facilities, several things are clear from the Bisconti data. First, there was a substantial increase in support for nuclear energy between 1983 and 2010 among those who live within 10 miles of a nuclear power plant, and that their 2011 data show a decrease in support of 4–9 % after a widely publicized event that had dramatic consequences. The Bisconti series are important for this study because of their geographical focus on sites immediately adjacent to nuclear power plants.

3.4.2.2 Pew, Gallup, and Harris Surveys

In September 2005 and March 2011, the Pew Research Center for the People & the Press (2011a, b) conducted multiple surveys on public preferences for different forms of energy. The first of these in 2005 showed 39 % favored the increased use of nuclear power and 53 % were opposed. The proportion who favored more nuclear power rose to 44 %, jumped to 50 %, back to 45 %, and up 52 %, and then back to 45 % in October 2010. The post-Fukushima proportion was 39 % based on a survey of 1,004 adult US residents on March 17–20, 2011. In other words, the post-Fukushima proportion is the same as the proportion in favor in September 2005. If the US public had really turned against nuclear power, I would have assumed a much more substantial decline immediately after the event occurred. The Pew reports show interesting demographic shifts. Men, persons 50+ years old, college graduates, and those identifying as Republican have been the strongest supporters. The strongest declines in support after Fukushima were among women, college graduates, and Republicans.

The Pew surveys also have interesting data about other energy sources. Most notably, in between September 2008 and February 2010, 63–67 % of respondents favored more offshore drilling for oil and gas. The Gulf oil blowout event occurred in 2010 and flow continued for about 3 months. By June 2010, the proportion fell to 44 %. But as of March 2011, it had risen back to 57 %. Also the Pew surveys showed relatively little change (there was a slight decline) in the strong support for federal support of research about wind, solar, and hydrogen technologies, for contributing more resources for mass transit, and for providing tax incentives for those purchasing hybrid motor vehicles.

Gallup has polled US residents about nuclear energy since 1994. In March 2010, Gallup (2011a,b) reported that "U.S. support for nuclear power climbs to a new high of 62 %." That is, 62 % of respondents favored or somewhat favored the use of nuclear energy to provide electricity in the USA. In 2001, the proportion was 48 %, and climbed. On March 15, 2011, the answer was 57 % favored, a decrease from 62 % in 2010 (Jones 2011; Gallup 2011a,b). Yet the 57 % was higher than their numbers in many previous years. The Gallup 2011 survey showed that 39 % were a "lot more concerned" about a nuclear disaster in the USA after the events in Japan, but 27 % were not more concerned. Also, 47 % opposed the construction of new nuclear plants in the USA, and yet 44 % favored them. Lastly, 58 % said they think nuclear power plants in the USA are safe compared to 36 % who said that they were not (Gallup 2011a,b).

Harris Poll Interactive also periodically polls US residents about energy. On March 23–25, Harris (2011), in their survey of 2,090 adults found that 41 % supported new nuclear plants in the USA, 39 % were opposed, and 20 % were uncertain. This compared to 49 % pro and 32 % con 3 years earlier. The poll also showed that 73 % of respondents feel that nuclear waste disposal is a "serious problem."

Overall, the Pew, Gallup, and Harris poll results suggest that the long-term impact of the Fukushima events on public preferences in the USA is not yet clear. There was a clear public split prior to Fukushima and the split continues, albeit more people are more negative than prior to the event.

3.4.2.3   US University-Based Surveys

A multidisciplinary group of MIT researchers has been comprehensively studying US energy choices, including economic, technology, and public policy dimensions. For example, in a 2007 sample of US residents, Ansolabehere (2007) found that 19 % wanted more reliance on coal and 36 % more reliance on nuclear. In the year 2002 study with the same questions, the fractions were 17 and 28 %, respectively. In contrast, over 90 % want greater reliance on solar and wind. They noted more favorable public attitudes about nuclear power.

Ansolabehere and Konisky (2009) addressed public attitudes toward construction of new power plants. They assert that there is a common thread running through public preferences and perceptions about siting, which is that perceived harm is the most important among the decision-making variables, making it difficult to site new nuclear, coal, and other facilities. Greenberg (2009a) has asserted that in response industry has adopted a CLAMP (choose locations at major plants) policy for not only nuclear but other energy facilities.

A major limitation of the literature is that few surveys provide insights about any of the DOE-managed defense waste nuclear sites. Accordingly, Jenkins-Smith et al. (2011) study of WIPP is helpful. The authors examined 35 New Mexico-wide public surveys from 1990 to the summer of 2001. WIPP accepted the first delivery of radioactive waste on March 26, 1999, after much heated debate within the state. In 1990 about 40 % favored opening the site, and this proportion did not change much until 1995 when it jumped to almost 50 %. When the site opened, support for opening was almost 60 %. Not surprisingly, perceived risk fell on a 5-point scale from an average of 3.2–2.9 during this same period. The authors' provided multiple regression models about preference for opening WIPP. The strongest correlates were male, white, income and education, identification with the Republican party, trust in performance of the New Mexico government, approval of EPA and DOE, and residence near the site. All of these are consistent with expectations as described earlier.

3.5 Post-TMI and Chernobyl Surveys

A logical place to gain insights about the lasting impact of the Fukushima events on public preferences and perceptions about nuclear power and waste management are public opinion polls following the Three Mile Island and Chernobyl events. Rosa and Dunlap (1994) captured the essence of the impacts in the USA in a paper published in *Public Opinion Quarterly* using four major polling groups. Beginning with the Louis Harris polls, between March 1975 and September 1978, an average of 60 % of respondents favored building new nuclear power plants in the USA, whereas opposition averaged about 22 %. Immediately after the event and for about a year, opposition grew to 43 % and support fell to 43 %. Both opposition and support fluctuated between 45 and 55 % until 1981 when opposition gradually increased.

Chernobyl occurred on April 26, 1986. Opposition during the next 4 years increased to over 65 % and support fell to less than 30 %. The polls show marked drops of about 10 % in support immediately after the events and increases in opposition. However, the authors question the so-called rebound hypothesis, which is that public support rebounds after a period of time. Their analysis of published data up through the early 1990s showed growing opposition to new nuclear power plants.

While not contributing to measurement of public response to nuclear events, Berry, Jones, and Powers (1999) reviewed the role of the mass media in four emergency situations. The authors report that despite conflicting information, incomprehensible technical information, unfounded rumors, and some sensationalized reporting, the public's response was not to panic. Rather those who were extremely concerned evacuated and others went on with their lives.

During both events, the nuclear industry was trying to understand public preferences and perceptions. General Public Utilities (1979) reviewed the TMI incident, technical background, and the social issues including public opinion. Their interpretation of TMI was not notably different from many others. For example, they noted a poll conducted by a local college of residents within 50 miles of TMI. That poll of 375 people showed that 62 % support nuclear power and that even 58 % favored continuing to operate the second nuclear unit at the site. However, they also reported national polls that showed an increase of about 10 % in public opposition to more nuclear power plants especially in the Northeast. They observed ambivalence among the public and cited a New York Times/CBS news poll that showed a drop support for new development of nuclear energy to 46 % after the incident from 69 % 2 years earlier. But they also cited as evidence of ambivalence the fact that the public by a margin of 2–1 favored nuclear power over importing oil. In addition, they cited from a *Washington Post* poll that showed an increase in opposition to new nuclear plants of more than 20 %, and a Gallup national poll that reported 45 % would be willing to live near nuclear plant, which was a decrease from 62 % 3 years earlier. It is notable that they (General Public Utilities 1979, p.14) concluded their discussion of public opinion with a quote from a Gallup survey as follows: "despite these worries, the American people are not ready to reject the use of nuclear power for future energy needs."

The Chernobyl incident in 1986 had a much greater impact on Europe than on the USA. A number of studies found a difference between people living in countries that received higher doses than those that did not (Renn 1990). Perceptions of nuclear power became much more negative in those countries with higher exposure levels.

Siegrist pointed out three longitudinal studies to this author. Two in Europe after Chernobyl and one in the USA followed small samples (69–154) of people before and after the incidents. Verplanken (1989) observed that perceptions of nuclear power were much more negative after the Chernobyl accident. However, they returned to almost pre-Chernobyl levels 6 months after the accident, and yet 19 months later had once again become more negative. The author hypothesizes that there was considerable negative media coverage of nuclear power during this period which may have influenced public perception. Eiser et al. (1989) compared pre- and post-Chernobyl perceptions about nuclear power facilities with those of coal and oil, as well as other industrial facilities. While acceptance of the other facilities did not change, acceptance of nuclear power plants dropped considerably. The third study, by Lindell and Perry (1990), found relatively little change in perception and acceptance of nuclear power.

The final survey reviewed in this chapter was a study of the impact of the Fukushima events on public preferences and risk beliefs in 24 countries by Ipsos (2011), an international research organization. The organization sampled 18,787 adults between May 6 and May 21, 2011, using an online panel. Approximately 1,000 people participated in most of the countries but in several the numbers were closer to 500, and the authors weighted the data so that it reflected national demographic characteristics. For context, the authors show that over 90 % of all respondents support solar, wind, and hydroelectric power, 80 % favored natural gas, 48 % coal, and 38 % nuclear energy. Only three countries, including India and the USA, showed a majority strongly or somewhat supporting nuclear power. Overall 34 % strongly opposed nuclear power and 28 % somewhat opposed it. The survey showed that 26 % of those who opposed nuclear power had been significantly influenced by the Fukushima events, with proportions over 50 % reported in South Korea, Japan, China, and India. By comparison, the USA reported that 26 % who opposed nuclear power were strongly influenced by the Fukushima events, and in the UK, the proportion was 20 %.

Trust, I believe, is a key issue, and consequently, it is noteworthy that only 54 % felt that Japanese officials and government leaders honestly communicated the nature and impact of the event to the Japanese people and others. Notably, only 17 % of South Koreans and 28 % of Japanese respondents agreed with that assertion, and the proportion of the USA was 56 %.

Globally, only 31 % of respondents wanted to continue to build nuclear energy capacity. However, the proportion was 44 % in the USA and 43 % in Great Britain, which were among the highest. In India, China, and Russia, the proportions supporting the building of new nuclear energy capacity were 49, 38, and 34 %, respectively.

Overall, the TMI and Chernobyl experiences were that nuclear power became less acceptable to the public after the events than before the events. How much less acceptable is the key question. Several of the surveys showed decreases of about 10 % in the immediate aftermath and several showed much larger decreases of 20–30 %, depending upon the location. The second key question was how long did the increase in opposition last? The Rosa and Dunlap (1994) paper concludes that there was a rebound and yet opposition resurfaced, and accordingly by early 1990 public opposition was much stronger than immediately after the incidents in Pennsylvania and the Ukraine.

What is fascinating and I think critical is that shortly after the publication of Rosa and Dunlap's excellent paper, the trends changed and support started to increase. No one knows exactly why. Accordingly, it is essential to consider the world energy and political environments in the present tense rather than to assume the kinds of changes in public preferences and perceptions that were observed after TMI and Chernobyl would occur after Fukushima. First and perhaps foremost, the Cold War has ended, and with it, the repeated imaging of nuclear weapons, nuclear weapons detonation, and the mushroom cloud have somewhat faded. Second, world consumption of energy continues to grow rapidly, and the USA and the EU are now competing with China, India, Russia, and other rapidly developing nations for what appears to be more limited or at least more expensive fossil fuel resources. Third, the USA, Europe, and many other countries are reluctant to permit their national economy and quality of life to be held hostage by international political events. In this regard, nuclear power can be perceived as a controllable energy asset. Fourth, fossil fuel plants, particularly coal, may be cheaper to build, but their association with global warming has contributed to increasing public support for renewable energy sources, and some consider nuclear energy a green renewable source and a lesser evil than global warming associated with fossil fuels (European Commission 2008; Roth et al. 2009; Visschers et al. 2011). Have these asserted benefits of nuclear power overcome the negatives? Or have the Fukushima events reignited the negative side of the nuclear industry for the public? Apropos the last point, the event occurred and images were immediately available not only on the traditional mass media but on computer screens all over the world. The potential for powerful negative emotional connections to nuclear power and waste management was amplified by the earthquake, tsunami, and then powerful images of the explosions.

The Fukushima events had to have been worrisome and painful for everyone associated with the nuclear industry and for residents who live near facilities. The public response to Fukushima is important globally, nationally, and locally and particularly important to the areas that host DOE's defense waste legacy. That waste legacy not only cannot be abandoned, but in addition, key members of the DOE believe that their sites can be used to improve a host of energy-related technologies, to generate more electricity, and to create local jobs. With or without a nuclear power renaissance, US federal government must manage and oversee these DOE and commercial sites. In 2010, it appeared that in the absence of a new

TMI or Chernobyl that large parts of the population had been becoming more open-minded about considering it because of potential benefits. Chapter 5 will explore how Fukushima has changed perceptions focusing on six key DOE waste management regions and including a national sample for comparison.

References

Alhakami A, Slovic P (1994) A psychological study of the inverse relationship between perceived risk and proceeds benefit. Risk Anal 14:1085–1096

Ansolabehere S (2007) Public attitudes toward America's energy options: insights for nuclear energy. MIT-NES-TR-08

Ansolabehere S, Konisky D (2009) Public attitudes toward construction of new power plants. Public Opin Quart 73:566–577

Berry L, Jones A, Powers T (1999) Media interaction with the public in emergency situations: four case studies. Library of Congress, Washington, DC

Bisconti AS (2007) Perspective on public opinion (December 2007). A report prepared for the National Energy Institute. Retrieved September 18, 2009 from http://nei.org

Bisconti AS (2008) Perspective on public opinion (November 2008). A report prepared for the National Energy Institute. Retrieved September 18, 2009 from http://nei.org

Bisconti A S (2009). Perspective on public opinion (June 2009). A report prepared for the National Energy Institute. Retrieved February 17, 2011, from http://nei.org/resourcesandstats/documentlibrary/publications/perspectiveonpublicopinion/june2009/

Bisconti A S (2010a) Perspective on public opinion (June 2010). A report prepared for the Nuclear Energy Institute. Retrieved February 17, 2011, from http://nei.org/resourcesandstats/document library/publications/perspectiveonpublicopinion/perspective-on-public-opinion-june-2010/

Bisconti AS (2010b) Public opinion snapshot (Winter, 2010). A report prepared for the Nuclear Energy Institute. Retrieved February 17, 2011, from http://nei.org/resourcesandstats/documentlibrary/publications/perspectiveonpublicopinion/winter-2010/

Bisconti Research Inc. (2011) National questionnaire for nuclear plant neighbor survey. June. http://www.bisconti.com. Accessed September 20, 2011

Bronfman N, Cifuentes L (2003) Risk perception in a developing country: the case of Chile. Risk Anal 23:171–185

Cha YJ (2000) Risk perception in Korea: a comparison with Japan and the United States. J Risk Res 3:321–332

Corner A, Venables D, Spence A, Poortinga W, Demski C, Pidgeon N (2011) Nuclear power, climate change and energy security: exploring British public attitudes. Energy Pol 39:4823–4833

Dickert S, Slovic P (2009) Attentional mechanisms in the generation of sympathy. Judge Dec Mak 4:297–306

Douglas M (1970) Natural symbols: explorations in cosmology. Barrie & Rockliff, London

Earle TC (2010) Trust in risk management: a model-based review of empirical research. Risk Anal 30:541–574

Earle T, Cvetkovich G (1995) Social trust, towards a cosmopolitan society. Praeger, London

Eiser JR, Spears R, Webley P (1989) Nuclear attitudes before and after Chernobyl: change and judgment. J Appl Soc Psychol 19:689–700

European Commission (2005) Radioactive Waste. Brussels: TNS Opinion & Social Report No.: Special Eurobarometer 227

European Commission (2006) Energy Technologies. Brussels: TNS Opinion & Social Report No.: EUR22396

European Commission (2008) Attitudes toward radioactive waste. brussels: TNS Opinion & Social Report No.: Special Eurobarometer 297

European Commission (2010) Europeans and nuclear safety. Brussels: TNS Opinion & Social Report No.: Special Eurobarometer 324

Finucane M, Alhakami A, Slovic P, Johnson S (2000a) The affect heuristic in judgments of risks and benefits. J Behav Dec Mak 13:1–17

Finucane M, Slovic P, Mertz CK, Satterfield T (2000b) Gender, race, and perceived risk: the 'White Male' affect. Health, Risk & Society 14:159–172

Finucane M, Slovic P, Mertz CK (2000c) Public perception of the risk of blood transfusion. Transfusion 40:1017–1022

Flynn J, Slovic P, Mertz CK (1994) Gender, race, and perception of environmental health risks. Risk Anal 14:1101–1108

Gallup (2011a) Majority of Americans say nuclear power plants in the U.S. are safe. http://www. gallup.com/poll/146939. Accessed September 23, 2011

Gallup (2011b) US support for nuclear power climbs to new high of 62 %. http://www.gallup.com/ poll/126827. Accessed September 23, 2011

General Public Utilities Corporation (1979) The TMI 2 Story. www.threemileisland.org/ dowjones/225.pdf. Accessed September 28, 2011

Greenberg M (2005) Concern about environmental pollution: how much difference do race and ethnicity make? a New Jersey case study. Environ Health Perspect 113:369–374

Greenberg M (2009a) NIMBY, CLAMP and the location of new nuclear-related facilities: U.S. National and eleven site-specific surveys. Risk Anal 29:1242–1254

Greenberg M (2009b) Energy sources, public policy, and public preferences: analysis of US national and site-specific data. Energy Policy 37:3242–3249

Greenberg M (2012) Comment: the affect heuristic, correspondence analysis, and understanding LULUs. Risk Anal 32:478–80

Greenberg M, Schneider D (1995) Gender differences in risk perception: effects differ in stressed vs. non-stressed environments. Risk Anal 15:503–511

Greenberg M, Truelove H (2010) Energy choices and perceived risks: Is it just global warming and fear of a nuclear power plant accident? Risk Anal 32:819–831

Greenberg M, Lowrie K, Krueckeberg D, Mayer H, Simon D (1997) Bombs and butterflies: a case study of the challenges of post cold-war environmental planning and management for the United States' nuclear weapons sites. J Environ Plann Manage 40:739–750

Greenberg M, Popper F, Truelove H (2012) Are LULUs Still Enduringly Objectionable? Environ Plann Manage 55(6):713–731

Gross J, Rayner S (1985) Measuring culture: a paradigm for the analysis of social organization. Columbia University Press, New York

Harris Poll Interactive (2011) Harris Poll about nuclear power. http://www.se-ygn.org/2011/03/31. Accessed March 23, 2011

Hsee C (1996) Elastic justification: how unjustifiee factors influence judgments. Organ Behav Human Dec Process 66:122–129

Hsee C, Kunreuther H (2000) The affection effect and insurance decisions. J Risk Uncertain 20:141–159

Ipsos (2011) Global citizen reaction to the Fukushima nuclear plant disaster. PowerPoint presentation. http://www.ipsos-mori.com/rsearchpublciations/rsearcharchive/2817. Accessed September 29, 2011

Jenkins-Smith H, Silva C, Nowlin M, deLozier G (2011) Reversing nuclear opposition: evolving public acceptance of a permanent nuclear waste disposal facility. Risk Anal 31:629–644

Jones J (2011) Disaster in Japan raises nuclear concerns in U.S. http://www.gallup.com/poll/ 146660. Accessed September 23, 2011.

Kahan D, Braman D, Gastil J, Slovic P, Mertz CK (2007) Culture and identity-protective cognition: explaining the white-male effect in risk perception. J Empirical Legal Stud 4:465–505

Keller C, Visschers V, Siegrist M (2012) Affective imagery and acceptance of replacing nuclear power plants. Risk Anal 32:464–477

Knight A (2007) Intervening effects of knowledge, morality, trust and benefits on support for animal and plant biotechnology applications. Risk Anal 27:1553–1563

Koren G, Klein N (1991) Bias against negative studies in newspaper reports of medical research. J Am Med Assoc 266:1824–1826

Kunreuther H, Easterling H, Desvousges W, Slovic P (1990) Public attitudes toward citing a high-level nuclear waste repository in Nevada. Risk Anal 10:469–484

Lindell M, Perry R (1990) Effects of the Chernobyl accident on public perceptions of nuclear plant accident risks. Risk Anal 10:393–399

Macfarlane M, Ewing R (eds) (2006) Uncertainty underground: yucca mountain and the Nations High-Level Nuclear Waste. MIT Press, Cambridge, MA

Nye J, Zelikow P, King D (1997) Why people don't trust government. Harvard University Press, Cambridge, MA

Olofsson A, Rashid S (2011) The white (male) effect and risk perception: can equality make a difference? Risk Anal 31:1016–1032

Parkhill K, Pidgeon N, Henwood N, Simmons P, Venables D (2010) From the familiar to the extraordinary: local residents' perception of risk when living with nuclear power in the UK. Transact Inst Br Geogr 35:39–58

Parkhill K, Henwood N, Pidgeon N, Simmons P (2011) Humour, affect and emotion work in communities living with nuclear risk? Br J Sociol 62:324–346

Peters E, Slovic P (1996) The role of affect and worldviews as orienting dispositions in the perception and acceptance of nuclear power. J Appl Soc Psychol 26:1427–1453

Pew Research Center (1998) Deconstructing distrust: Americans view government. Pew Research Center, Washington DC

Pew Research Center for the People & the Press (2011a) Opposition to nuclear power rises amid Japanese crisis. Http://People-press.org/2011/03/21. Accessed September 19, 2011

Pew Research Center for the People & the Press (2011b) Japanese resilient, but see economic challenges ahead. Http://People-press.org/2011/06/01. Accessed September 19, 2011

Pidgeon N, Lorenzoni I, Poortinga W (2008) Climate change or nuclear power–no thanks! A quantitative study of public perceptions and risk framing in Britain. Global Environ Change 18:69–85

Poortinga W, Pidgeon N (2003) Exploring the dimensionality of trust in risk regulation. Risk Anal 23:961–972

Rayner S (1992) Cultural theory and risk analysis. In: Kirmsky S, Goldin D (eds) Social theories of risk. Praeger, Westport, CT, p 83

Renn O (1990) Public responses to the Chernobyl accident. J Environ Psychol 19:151–167

Rivers L, Arvai J, Slovic P (2010) Beyond a simple case of black and white: searching for the white male effect in the African-American community. Risk Anal 30:65–77

Rosa E, Dunlap R (1994) The polls-poll trends, nuclear power: three decades of public opinion. Publ Opin Quart 58:295–325

Roth S, Hirschberg S, Bauer C et al (2009) Sustainability of electricity supply technology portfolio. Ann Nucl Energy 36:409–416

Salvadori L, Savio S, Nicotra E, Rumiati R, Finucane M, Slovic P (2004) Expert and public perception of risk from biotechnology. Risk Anal 24:1289–1299

Siegrist M (2010) Trust and confidence: the difficulties in distinguishing the two concepts in research. Risk Anal 30:1022–1024

Siegrist M, Cvetkovich G (2000) Perception of hazards: the role of social trust and knowledge. Risk Anal 20:713–719

Siegrist M, Cvetkovich G (2001) Better negative than positive? Evidence of a bias for negative information about possible health dangers. Risk Anal 21:199–206

Siegrist M, Cvetkovich G, Roth C (2000) Salient value similarity, social trust, and risk/benefit perception. Risk Anal 20:353–362

Sjoberg L (2002) Limits of knowledge and the limited importance of trust. Risk Anal 21:189–98

Sjöberg L (2003) Risk perception is not what it seems: the psychometric paradigm revisited. In: Andersson K (ed) VALDOR Conference 2003. VALDOR, Stockholm, pp 14–29

Skowronski J, Carlston D (1989) Negativity and extremity biases in impression formation: a review of explanations. Psychol Bull 105:131–142

Slovic P (1987) Perception of risk. Science 236:280–285

Slovic P (2001) Cigarette smokers: rational actors or rational fools? In: Slovic P (ed) Smoking: risk, perception and policy. Sage, Thousand Oaks, CA, pp 97–124

Slovic P, Finucane M, Peters E, MacGregor D (2004) Risk as analysis and risk as feelings: some thoughts about affect, reason, risk, and rationality. Risk Anal 24:311–322

Slovic P, Peters E, Grana J, Berger S, Dieck GS (2007) Risk perception of prescription drugs: results of a national survey. Drug Inform J 41:81–100

Small D, Loewenstein G (2003) Helping the victim or helping a victim: altruism and identifiability. J Risk Uncertain 26:5–16

Stern P (2000) Toward a coherent theory of environmentally significant behavior. the 'White Male' effect. Health, Risk & Society 2:159–172

Vandermoere F (2008) Hazard perception, risk perception, and the need for decontamination by residents exposed to soil pollution: the role of sustainability and the limits of expert knowledge. Risk Anal 28:387–398

Vastfjall C, Peters C, Slovic P (2008) Affect, risk perception and future optimism after the tsunami disaster. Judge Dec Mak 3:64–72

Verplanken B (1989) Beliefs, attitudes, and intentions toward nuclear energy before and after Chernobyl and a longitudinal within-subjects design. Environ Behav 21:371–392

Visschers VHM, Keller C, Siegrist M (2011) Climate change benefits and energy supply benefits as determinants of acceptance of nuclear power stations: investigating an explanatory model. Energy Policy 39:3621–3629

Visschers VHM, Siegrist M (2012) How a nuclear power plant accident influences acceptance: results of a longitudinal study before and after the Fukushima disaster. Risk Anal

Weart S (1992) Fears, fantasies and fallout. New Scientist 136:34–37

Xie X, Wang M, Xu L (2003) What risks are Chinese people concerned about? Risk Anal 23:685–695

Chapter 4
CRESP Surveys of Major US Department of Energy Environmental Management Site Regions and of the National Population, 2005–2010

Abstract CRESP conducted surveys of areas near major US DOE sites in 2005, 2008, 2009, 2010, and 2011. The 2008–2011 surveys also collected a national sample. This chapter describes the survey questions, the protocols, and the results of the 2005, 2008, 2009, and 2010 surveys. Two-thirds of site-specific sample favors DOE's energy park concept, with 36 % favoring it for their own area, and site-specific support for new waste management activities was >50 % and was higher than national sample. Respondents preferred environmental and risk-based management policies that monitor the water and air at the site, workers, and strongly supported equipment and training for local first responders and tools to alert the public. Requiring site managers to live near sites, prohibiting new missions, providing guided tours of the sites, and other organizational steps were the least favored priorities. Emotions and feelings were the strongest correlates, especially with regard to environmental management options. The site-specific sample had many more positive feelings, emotions, and images than the national sample and many more links to positive economic outcomes. Also, affluent college-educated white males disproportionately supported new on-site activities as did those who were optimistic about the future.

4.1 Introduction

CRESP surveys in 2005, 2008, 2009, and 2010 were focused on answering three questions:

1. How receptive are people who live near nuclear waste management facilities to new plants and activities, and also to power plants, laboratories, and other nuclear-related facilities? This is the future nuclear use question.
2. What are their preferences for different waste management policies at existing US DOE waste management sites? This is the legacy waste management question.

M.R. Greenberg, *Nuclear Waste Management, Nuclear Power and Energy Choices*,   65
Lecture Notes in Energy 2, DOI 10.1007/978-1-4471-4231-7_4,
© Springer-Verlag London 2013

3. What emotions and images, feelings of trust about authorities, demographic attributes, cultural beliefs and values, and personal history and preferences are associated with public responses to the first two questions? This is the correlates question.

The results summarized in this chapter capture pre-Fukushima answers to these three questions. We believe that the answers pre- and post-Fukushima are of practical significance to the US Department of Energy as well as to the US Nuclear Regulatory Commission; the US Environmental Protection Agency; other federal, state, and local agencies that have an added responsibility to nearby residents for oversight of the facilities; and contractors who manage the sites. As noted in Chap. 3, few surveys have focused on nuclear waste management, especially the defense-waste legacy, and there is virtually no data available that would help the DOE understand the nearby public's views of possible new operations on its sites (future nuclear use question) and what the public would like it to do to protect public health and the environment (legacy waste management question). Furthermore, it is important that we try to understand characteristics that are associated with the range of public perceptions and preferences about nuclear waste management (correlates question).

Secondary objectives of the surveys not directly addressed in this chapter were (1) to assess preferences for different energy sources and conservation; (2) examine the relationship between survey results, media coverage of the issues, and DOE site-specific advisory board deliberations; and (3) measure public knowledge of energy facts such as how much we rely on different energy sources, where the sources are located and where spent fuel is managed. The findings of the CRESP 2005, 2008, 2009, and 2010 surveys, but not the 2011 one, which is the focus of Chap. 5, have been presented in more than a dozen publications (Greenberg et al. 2007a, b, c, 2008, 2011a, b; Greenberg 2009a, b, c, 2010 ; Lowrie et al. 2000; Lowrie and Greenberg 2000, 2001; Greenberg and Truelove 2010, 2011; Greenberg and Lowrie 2002). This chapter will not repeat all the findings; rather, it will illustrate key findings from these studies and add more detail about the survey process, which was precluded by space limitations in the papers.

The chapter is divided into five parts. It begins with a summary of survey locations (see Chap. 2 for a more detailed presentation and maps), followed by a discussion of the survey design, the questions, and then a selective summary of the major results and a discussion of their implications.

4.2 Locations

Every survey included the DOE's Hanford, Idaho, Oak Ridge, and Savannah River sites, which have been responsible for the major DOE cleanup expenditures. Several surveys included Fernald and Mound in Ohio and Rocky Flats in Colorado. The last three were included when these sites were in transition from cleanup to final closure. More recently, we have included Los Alamos and WIPP in the surveys. Hanford, Idaho, Los Alamos, Oak Ridge, Savannah River, and WIPP have major nuclear hazardous waste legacies. Reiterating a fact from Chap. 2, their aggregate budget for waste management has averaged about 75 % of the DOE's environmental management budget (Greenberg et al. 2011). Beginning in 2008, the surveys asked almost identical questions of a national sample in order to have a comparative perspective for the site-specific findings.

Again, with one exception, every survey defined a DOE-centered region as an area within 50 miles of one of the DOE sites. A 50-mile sample was a compromise that allowed us to look for a familiarity-based halo effect and also has a non-DOE affiliated sample. Many of the areas within 20 miles have had a strong economic relationship with the sites, and if the sample population was drawn only within 20 miles, we expected to find a bias toward less concern about the site. Yet, using limited sampling dollars to test beyond 50 miles was an inefficient use of resources because the overwhelming number of samples would be from unfamiliar people who would not be directly impacted by site activities. One survey, the exception, used a 100-mile radius in order to include locations with coal, gas, and other major energy facilities.

4.3 Survey Design

The ideal survey design to answer the three research questions includes conversations with prominent individuals, such as local elected officials, local citizen group leaders, a review of mass media coverage, and discussions with groups of local residents to determine what they believe to be important. In the case of the DOE site regions, we had minutes from DOE site-specific advisory boards (SSABs) (Greenberg et al. 2008; Lowrie et al. 2000; Lowrie and Greenberg 2000, 2001). The CRESP project began in 1995, and consequently, we had a decade to accumulate a great deal of information from interviews, conversations, reading minutes, and newspapers before we conducted a single survey. While not all of these pre-survey and ongoing efforts were published, all were used to construct survey questions that we believed would be relevant for DOE staff and other interested parties (Lowrie et al. 2000; Lowrie and Greenberg 2000, 2001; Greenberg and Lowrie 2002).

Before presenting the results, we describe the data collection method we chose and how trends in survey research influenced the design of these surveys. We had four survey options: (1) face-to-face interviews, (2) mail, (3) computer, and (4) telephone. Face-to-face interviews are productive if the surveyors are highly trained. They can begin with a baseline set of standardized fully labeled questions and then, depending upon the responses of the individual being interviewed, the surveyor can ask additional follow-up questions. Limitations of this approach are the requirement for highly skilled interviewers, difficulty of obtaining a representative sample, and high cost. We were especially concerned about replicability across the range of DOE sites spread out across the USA, that is, interviewer variation by site could bias the results. This approach was not used for these surveys but has been for others.

Questionnaires mailed to respondents were standard for many years. However, in a country where paper mail use is being replaced by electronic mail, response rates have plunged. While mail surveys are still used when targeting specific populations or neighborhoods, and to contact people to alert them that a phone or computer survey is coming, relatively few national major surveys use letters. Asking people to respond to questions delivered to them through a computer has rapidly become a standard approach, especially when the target audience can be identified from a directory. So, for example, if we wanted to know how local and state planners think about DOE sites in their states and local areas, we would communicate with them through the computer, at least initially. A computer design is more problematic for a random public survey because not everyone has equal access to a computer, is capable of responding to surveys on a computer, or is willing to do so. Poor and senior populations would be underrepresented. Affluence, education, and age are associated with public preferences and perceptions about many phenomena, including energy, and hence, this was not a viable approach for these surveys.

Our 2005–2010 surveys were administered over landline telephones using random digit dialing (RDD) and following American Association for Public Opinion Research (AAPOR 2009) standards. Our objective was to obtain a cooperation rate of 30 % and a response rate of 20 %. Given that not long ago the author expected a 50–60 % response rate with three phone calls to each respondent, it is legitimate to question the validity of the survey data. This change necessitates a substantive reply in this chapter that is directly relevant to nearly every RDD survey, not just this one.

The most important point is that there has been a dramatic decline in public willingness to respond to phone surveys (Zukin 2006; Cantor et al. 2009). In order to achieve our sampling targets, after eliminating bad numbers (such as nonresident, not in service), the plan was to call each good number 8–11 times.

Too often, the terms response and cooperation rates are not defined by authors, leading to confusion. We used the same definitions in all our surveys. A response rate is the number of complete interviews divided by the number of eligible respondents. While this sounds simple, in reality, there are multiple definitions of what goes into the numerator and the denominator. AAPOR provides six different ways of calculating a response rate. We define and illustrate what our team did.

We have used AAPOR3 in our calculations, which is a widely used standard:

$$\text{Response rate} = I/[(I + P) + (R + \text{NC} + O) + e(\text{UH} + \text{UO})], \quad (4.1)$$

where I is the number of completed interviews and "complete" does not necessarily mean that every question has been answered in every survey. It also includes screen-outs who are people who are willing to cooperate but do not qualify according to the standard for admission into the survey. In our surveys these would include people who are not at least 18 years old and those who want to respond to one of the site-specific surveys but do not live in the defined region. For our context, in our surveys, completes outnumber screen-outs 14–1. P is the number of partially completed interviews, where a partial interview is providing insufficient information to make the response useful to the analyst. Completes outnumber partials 13–1. R are refusals and break-offs, and these outnumber completes 2–1. NC marks people who could not be contacted, O represents other nonresponses that cannot be followed up, and e is the estimated eligibility of unknowns, which is a calculation that compares those classified as $I,P,R,NC,$ and O with the sum of the above plus the not eligible.

For the Savannah River site, the response rate to the year 2010 survey was 22.3 %, calculated as follows:

$$RR_{\text{Savannah River}} = 388/[(388 + 37) + (798 + 247 + 64) \\ + (0.239(581 + 290)] = 0.223, \tag{4.2}$$

where $e = (388 + 37 + 798 + 247 + 64)/[(388 + 37 + 798 + 247 + 64) +4,885] = 0.239$.

The cooperation rate we used is the commonly used AAPOR3, was 31.7 %, which is the proportion of all cases interviewed of all eligible possible respondents who were contacted:

$$\text{Cooperation rate} = I/(I + P) + R. \tag{4.3}$$

$$\text{Cooperation rate}_{\text{Savannah River}} = 388/(388 + 37) + 791) = 0.317. \tag{4.4}$$

In other words, the number of completed interviews is divided by the number of completed and partially completed interviews plus the number of refusals and disconnects. Overall, 7,290 phone numbers were used to produce 388 complete surveys, of which 350 were used. Please note that the largest category was the not eligible group; in other words, the largest number of calls was to landline phone numbers without an eligible respondent.

Summarizing, a response rate of 20 % means that through a random process, we were able to do a complete interview with 1 of 5 eligible respondents. A cooperation rate of 30 % means that we were able to interview 3 out of every 10 eligible respondents that were contacted.

Are the results less useful because of these response rates? Every major survey faces the same question, so there is considerable debate about the answer to this question. The intuitive assumption is that high response and cooperation rates mean more reliable data. That assumption has been questioned. First, survey experiments show that lower response and cooperation rates do not mean more biased data. Their observations come from study designs that asked the same questions with different numbers of callbacks and found similar results and from comparisons of demographic benchmarks from the US Census and other federal government surveys that have over a 90 % response rates (AAPOR 2008, 2009; Merkle and Edelman 2002; Keeter et al., 2000; Curtin et al. 2000; Public Opinion Quarterly 2006; Pew 2004). Second, some survey experts are more concerned about cooperation rates than response rates. They assert that refusals may imply a different view about the survey issues, a view shared by the first author (John F. Kennedy School of Government 2009). Overall, the response and cooperation rates reported here are consistent with the survey research literature at this time and reflect a marked change in survey research design and practice during the last decade.

The proliferation of cell phones is a major complication for survey researchers. A growing number of people, now estimated to be 30 %, use only a cell phone, an increase from 8 % in 2005 (CTIA 2011). A landline-only survey in the year 2011 and going forward faces substantial bias in responses. Our 2011 survey had a 25 % cell phone and 75 % landline design (see Chap. 5).

Our 2005–2010 landline-only surveys compared respondent characteristics by age, race/ethnicity, and location. We found, as expected, undersampling among young and nonwhite populations. Hence, every regional survey was weighed by age (18–44, 45–64, and 65+) and white–nonwhite, which reduced the bias. The weighting was region specific. That is, for example, the weights for the Savannah River region were drawn from the most recent US Census population analyses (for a 2010 survey, these were 2009), not from national census estimates (see illustration below in Table 4.10). The national samples were weighted by national age data, white/nonwhite, and regional location. However, another reality of survey sampling is that it is not possible to entirely correct with weighting because not all factors that influence results are weighted.

Sample size is always an important issue. It is essential that we recognize that there are diminishing returns in terms of lower sampling error as more samples are added. For example, the sampling error is 7.2 % for a sample of 190, 5.3 % for 350 samples, 4.1 % for 600, and 3.5 % for 800 completed responses. If 50 % of a sample of 800 favors a new energy park, then we can be 95 % certain that the actual number is between 46.5 % and 53.5 %. The site-specific samples ranged from 191 to 350 for each site, and the national sample sizes ranged from 600 to 800. These numbers show a diminishing return in sampling error with increasing numbers of samples. For example, 100 samples produce a sampling error of 9.8 %. Adding 500 more samples ($n = 600$) reduces the sampling error to 4.0 %. Yet adding another 500 samples ($n = 1,100$) further reduces the sampling error to 2.9 %, only 1.1 % more. How many samples you need depends upon desired confidence limits.

In every survey, there is a budget to be expended and that budget can be spent on more questions, more callbacks to increase the response and cooperation rates, and more samples. Every CRESP survey balanced those three elements, and the final decisions were informed by a pilot test that calculated time per survey and clarity of every question. With regard to length, it is critical to realize that long phone surveys lead to people breaking off and others not giving thoughtful answers. Experiments have shown that respondents begin to lose concentration after 25 min and will give the same answer to every question. Consequently, our target has been about 20 min.

We illustrate the trade-off between number of questions, callbacks, and samples with the year 2010 survey. We collected 2,751 surveys: 651 were national and 2,100 were site specific. The site specific were 350 at each of six DOE sites (Hanford, Idaho, Los Alamos, Oak Ridge, Savannah River, and WIPP). Given these sample sizes, the margin of sampling errors was 3.8 % for the national cross-sectional sample, 2.1 % for the aggregate of the site-specific surveys, and 5.2 % for each of the six sites. The average interview duration was 22.8 min for the national sample and 22.2 min for the site-specific ones. We were able to obtain the desired response and cooperation rates with a 9 or 10 callback design. If we wanted to have a larger sample, then we would have either been forced to cut questions or number of surveys to stay within the budget.

Decisions about several other key sampling issues are presented here to complete the record. Survey results can be influenced by when the survey is conducted, especially if it is taken at different times of the year when similar respondents may not be available. We attempted to manage this possible problem by conducting the field surveys beginning during the first week of July and ending on the last day of August of every year.

Specific events can markedly influence results, for example, the Fukushima event. Other surveys have been conducted shortly after the Fukushima event, and many of these are described in Chap. 3. We deliberately chose to hold back our survey to our regular July–August slot in order to let some time pass and give us the opportunity to compare the results of our 2011 survey with those of others and with our own previous surveys. There is no absolute right or wrong about the timing issue, except for the need to try to be as consistent as possible and to ask a specific question or questions to isolate any event affect.

A case in hand was the impact of the major coal impoundment failure in Kingston, Tennessee, in 2008. During the administration of our 2009 survey, which had some questions about coal, the US Environmental Protection Agency released a list of high-risk coal waste impoundments that received media attention. Anticipating that some people would remember the original event, we had questions about the earlier event and wanted to see how these memories were related to preference of coal as a source of electrical energy. The survey was stopped for six days, and we added a question to determine if the EPA release influenced the results. Over 2,500 respondents were asked these coal questions. A total of 37 % knew about the impoundment break, but few had seen the EPA press release. A much larger proportion of those who were aware of the impoundment failure which flooded a valley, leading to massive ecological damage and considerable property damage, were opposed to relying more on coal use to generate electricity in the USA than those who knew nothing about the event or the EPA release (Greenberg and Truelove 2010).

We close the design section with two final points. All of the surveys were pilot-tested. The author listened to the tests in order to determine if any of the questions were unclear, leading to modification of some questions. Finally, multilingual interviewers were available and about 2.5 % of the interviews were conducted in Spanish.

4.4   Questions

This section presents the questions used to answer the three research questions.

4.4.1   Future Nuclear Use

Future site use and reuse questions were asked in the 2005, 2008, 2009, 2010, and 2011 surveys. In 2008 we studied nuclear waste management, power, and technology facilities: "I'm going to read some options about locating new nuclear facilities in the USA. For each set of options, please tell me which one you prefer." The options were (1) building a new nuclear waste disposal facility at a site where one already exists, (2) at a location elsewhere, or (3) do not build any new disposal facilities in the USA. The same set of questions was repeated for nuclear power plants and then nuclear laboratory and research facilities. In other words, there were nine site-preference options.

In 2009 we asked preferences about building new "energy parks, defined as 'using some of its nuclear waste management and energy production sites to conduct energy research on generating energy through various means including solar, wind, nuclear, coal, biofuels, and other renewable energy sources'." This question was slightly modified in 2010, and the same question was asked in 2011 in order to determine the Fukushima affect, if any (see Chap. 5).

And in 2005, as part of a set of environmental and risk management questions, we asked about public reaction to allowing no new nuclear-related activities on the sites. The form of that question is described below as part of the risk management question set.

4.4.2   Legacy Waste Management

In 2005, we asked respondents to rate 17 environmental and risk management policy options, and in 2010, we asked them to respond to 10 options; 9 of the 10 were the same. The preface to the questions was:

In addition to removing materials from the site, some radioactive hazards are being left to decay in place so they will be less dangerous in the future. Some of these materials may remain at the site for many years. Please help us understand the steps that the Department of Energy could take that would bring you the greatest peace of mind about this site.

They were then read a list of options, and these were randomized to avoid order effects. A ten-point scale was used: 1 = very low priority, 10 = very high priority. The 2005 questions produced some results that clearly were not high priorities among respondents. Some of these were dropped and replaced. The nine common to both surveys are listed below:

(a) Regularly monitor the health of site workers
(b) Provide specialized training and equipment to emergency response personnel from surrounding areas
(c) Regularly monitor the health of people who live near sites
(d) Continuously sample the quality of the air and water at the site
(e) Install an early warning system to alert residents to any problems
(f) Require that the government report information about the site to community representatives on a regularly scheduled basis
(g) Provide information about what is happening at the site using web sites and visits by site personnel to schools and community groups
(h) Make sure that the site remains owned by the federal government until all the hazards are removed
(i) Maintain the role of Citizens Advisory Boards that currently represent community interests at the site

4.4.3   Correlates

Chapter 3 reviewed previous studies, and this literature is not repeated here. Instead we highlight some of the key theories as the context of our questions. The five conceptual areas are presented in order of our expectations about their importance for understanding variations in preferences for new site activities and environmental practices.

Emotional reaction to a facility and activity was expected, by far, to be the strongest correlates of public response. These reactions were measured by fully labeled Likert scale questions and by open-ended questions.

As illustrations, we list six of the 17 questions that were asked. These concerns emerged from discussions with stakeholders. Also note that during the survey, these were randomized to avoid any order effect.

The US Department of Energy manages the [surveyor names the facility] site where nuclear weapons components were developed, tested, or manufactured. It has removed some hazardous materials from the site and plans to remove more. I'm going to read you a list of activities related to this cleanup. Please tell me how much each worries you. Do you worry about that a great deal, some, not much, or not at all?

1. That the cleanup of chemical or radioactive materials at the site will expose residents to hazards
2. That fish, bird, and animal habitats will be destroyed during the cleanup of the site
3. That workers involved in the cleanup will be exposed to hazards
4. That disturbing corroding or leaking storage containers will cause more damage to the environment than leaving them in place
5. There may be accidents when hazardous waste materials are transported to and from the site
6. Area residents will lose jobs if the site reduces its cleanup efforts and other activities.

The response options were as follows: (1) a great deal, (2) some, (3) not much, and (4) not at all. Do not know and refused were also recorded.

After several years, we observed, not surprisingly, that the answers to many of these 17 questions were strongly correlated. That is, those who worried a great deal about the cleanup of chemical or radioactive material exposing people would disproportionately also be worried about many of the other possible negative consequences. Accordingly, we created a "site worry scale" by testing the set of worry indicators with Cronbach's alpha, a statistic that tests the reliability of a set of indicators as a single scale. A score of 0.7 or higher is desirable, and scores of 0.8 or 0.9 or more are even more desirable values to justify a scale. These targets were easily met.

The feelings attached to waste sites can also be measured with open-ended questions. In 2010, for example, we asked respondent to "please take a moment to think about the DOE site in your region." We then asked them to list an "emotion or feeling that comes to your mind that describes the site." Next, we asked about an "image or picture" that comes to mind. And third, we said "thinking about that DOE site, please tell me the first color that comes to mind." The surveyors were told to record the answers verbatim, and the author read and classified them.

Trust was the second set of theory-grounded questions. As described in chapter 3, the literature partitions trust into components of which "competence," "communications," and "values" are the most common. The key parties that must be trusted are the DOE, NRC, and contractors. Yet, EPA, the Department of the Interior that works on some sites, other federal agencies, state, and even local government may have roles, as well as experts and the mass media. We have tested a variety of agency groupings. We found that the vast majority of the public cannot separate the DOE, the NRC, and other federal agencies, and the state and local agencies were not consistently strongly predictive nor were the experts and media. Consequently, the key trust questions and preamble being used in recent surveys are as follows:

> Now I'd like to get your general opinion of the government and private owners and managers. For each statement I read, please tell me whether you strongly agree, agree, disagree, or strongly disagree.
>
> 1. The state and federal government will make sure that underground materials at the site do not pollute the air, land, and water outside of the site's boundaries.
> 2. Site contractors will make sure that materials at the site do not pollute the air, land, and water outside of the site's boundaries.
> 3. The state and federal government communicates honestly with the people in this area.
> 4. Site contractors communicate honestly with the people in this area.
> 5. I trust the state and federal government to effectively manage any new activities.
> 6. I trust contractors to effectively manage any new activities.

The response options were (1) strongly agree, (2) agree, (3) disagree, and (4) strongly disagree. Do not know and refused were also recorded.

Demographic characteristics were the third indicator set. The initial assumption is that there would be a white male effect, that is, the group with the greatest access to power and the economic system would disproportionately favor new activities and be less demanding of environmental management practices. Measures of race/ ethnicity, income, education completed, age, and respondent sex were included.

Cultural beliefs and worldviews were the fourth set. As noted earlier, some researchers have challenged the demographic characteristics as indicators of culture beliefs and values. We have tested a number of these belief and value questions, and they form the fourth set of questions. After initial testing, the following six have been used for three surveys.

For each statement I'm about to read, please tell me whether you strongly agree, somewhat agree, somewhat disagree, or strongly disagree.

1. Our society would be better off if the distribution of wealth was more equal.
2. A lot of problems in our society come from the decline in the traditional family, where the man works and the woman stays home.
3. Discrimination against minorities is still a very serious problem in our country.
4. The government interferes far too much in our everyday lives.
5. I feel that people who are successful in business have a right to enjoy their wealth as they see fit.
6. Too many people expect society to do things for them that they should be doing for themselves.

These questions were scaled as follows: (1) strongly agree, (2) somewhat agree, (3) neither agree nor disagree (voluntary only), (4) somewhat disagree, and (5) strongly disagree. Again do not know and refused were recorded.

Personal history and associated preferences was the fifth set of indicators. Experience leads some people to be more comfortable with nuclear waste management and technology than others. We asked respondents to tell us about their personal familiarity with their local DOE waste management site, their knowledge and preferences about nuclear and other electrical energy sources, their personal involvement in environmental groups, and their feelings about the quality of the local environment in their area now and in 25 years. Some of these personal histories are not independent of age and political party affiliation. Some of the questions were fill in a blank, others provided several options, and still others were standard five-point Likert scales. Rather than provide the format of every questions, those that have added to our understanding are described in greater detail below in the results section.

Before turning to the results, we note that the purpose is not to repeat all the details of the papers we have published from these surveys. Rather it is to provide context for the next chapter, which includes the Fukushima events. Accordingly, large tables are adapted to highlight certain findings, and statistical significance tests are only produced for key findings.

4.5 Results

4.5.1 Question 1. Future Nuclear Use

Tables 4.1–4.3 summarize key findings from our questions about new site activities. "CLAMP" is an acronym for choosing locations at major plant site and means adding new activities to existing sites rather than other locations. "New" in the table

Table 4.1 Public preference for concentrating locations at major plants (CLAMP), 2008

Site	Nuclear power, % favor			Waste management, % favor			Laboratory, % favor		
	CLAMP	New	None	CLAMP	New	None	CLAMP	New	None
National sample (n = 600)	34	31	35	52[a]	18	30	50	22	28
Six DOE sites (n = 1,146)	35	31	34	56	18	26	51	20	29
Idaho site (n = 191)	47	41	12	67	21	11	68	23	10
Respondent or family member worked at the site (n = 398)	50	29	21	62	20	18	61	22	16

Source. Adapted from Greenberg (2009a), Table 1
CLAMP stands for choose locations at major plants
[a]The national sample value was significantly lower than the site-specific one at $p < 0.10$ and the Idaho and familiarity one at $p < 0.05$

Table 4.2 Public preference for energy parks, %, 2009

Site	Favor and in my state	Favor but in another state	Favor and no preference for site	Neutral	Against
National sample (n = 800)	29[a]	11	26	22	12
Six nuclear sites (n = 2,400)	36	10	22	22	10
Respondent or family member worked at the site (n = 302)	54	8	15	17	6

Source. Adapted from Greenberg 2010a, Table 1
[a]The national sample value was significantly lower than the site-specific and respondent familiarity ones at $p < 0.05$

means placing the activity at an alternative location, and "none" means that the respondent does not want any new factories and activities in the USA.

Table 4.1 shows remarkable agreement between the national and the DOE site-specific samples about the policies for all three of facilities. Beginning with the national sample, about half in the 2008 sample would add new management and laboratory facilities at locations that already have them, and about a third would add new nuclear power plants in their area. The "none" designation was more prominent for nuclear power plants (35 %) than for waste management and laboratory facilities (28 %). The six site DOE aggregate was virtually identical for nuclear power plants and laboratory facilities. However, it was four percent higher (56 vs. 52 %) for new waste management facilities, a difference significant at $p < 0.10$. This suggests a halo effect, which was observed among the group of respondents who worked at the site or had a family member who did. Table 4.1 also showed the Idaho effect, which we have observed in every survey. Idaho respondents disproportionately supported new activities in their area.

Table 4.2 compares public preferences for an energy park concept that the DOE has been promoting (Greenberg 2010). The idea is to use available DOE land, infrastructure, as well as the technical and financial analyses capability of its

Table 4.3 Rating of a policy that would prohibit new nuclear activity at nearby DOE nuclear site, 2005, %

Site	Six DOE sites (n = 1,351), %	Idaho site (n = 223), %	Rocky Flats site (n = 225), %
1 = very low priority	15	27	4
2–5	25	36	13
6–9	18	19	15
10 = very high priority	42	18	68

Source: Greenberg et al. (2007a)

employees to develop solar, wind, fuel cell, energy transmission, nuclear, coal, and other energy-related technologies that would improve the USA's energy supply and create jobs in these areas. Table 4.2 shows more support for the energy park concept at the six DOE than the nation as a whole and even more support in host counties.

The data in Table 4.3 were from 2005, a year when the US economy was expanding. Prohibiting new nuclear-related activity at a nearby DOE site was offered as one of 17 different environmental and risk management options. Six out of 10 respondents rated the policy a high priority (6–10) and 4 of 10 rated it a low one (1–5), although a score of 10 (highest priority) was three times as likely as a score of 1 (lowest priority). Table 4.3 also shows striking differences among the sites, illustrated by Idaho and Rocky Flats. Only 37 % of Idaho respondents would ban new nuclear facilities at their site compared to 83 % at Rocky Flats. The Idaho site has many DOE missions, sits on the high plains with a very small surrounding population, and was the location where the first commercial nuclear energy was generated. In strong contrast, downtown Denver was visible from the closed Rocky Flats in Colorado.

Summarizing, DOE's sites were notably more supportive of the energy park concept than their national sample counterparts. People who lived in the host counties and had personal ties to the site and residents of the Idaho site region were the most supportive.

4.5.2   Question 2. Legacy Waste Management

Seventeen management alternatives were presented to respondents in 2005 and 10 in 2010. We summarize the 2005 set because the 2010 results were remarkably similar, following the exact same rank order, and the average scores typically differed by 0.1–0.2 on a ten-point scale. Six were public health options (PH1-6), five were site restriction policies (SR1-5), and six were organizational alternatives (OR1-6) (Table 4.4). As noted above, the scale was 1–10, where 1 was the lowest priority and 10 was the highest. The results are summarized as the average value and the proportion that rated the alternative at the highest priority. Four of the five highest averages were public health surveillance of the air, water, and workers. It is

Table 4.4 Public preferences for environmental management practices, 2005

Option SR-site restrictions PH-public health surveillance ORG-organizational options ($n = 1{,}351$)	Average value (SDV) [range is 1 (low priority) to 10 (high priority)]	% selected the maximum value
PH-1: Continuously sample the quality of the air and water at the site	9.0 (1.9)	64
PH-2: Regularly monitor health of site workers	8.9 (2.0)	64
SR-1: Make sure that federal government owns site until all hazards are removed	8.8 (2.2)	65
PH-3: Provide specialized training and equipment to emergency response personnel from surrounding areas	8.6 (2.4)	58
PH-4: Install an early warning system to alert residents to any problems	8.5 (2.4)	60
ORG-1: Require government report information about site to community representatives on a regularly scheduled basis	8.4 (2.3)	53
SR-2: Maintain security around site by keeping visitors and recreational users off the site	8.3 (2.5)	54
PH-5: Regularly monitor the health of birds, fish, animals, and plants at the site	8.2 (2.3)	47
PH-6: Regularly monitor the health of people who live near the sites	8.2 (2.4)	49
ORG-2: Create a trust fund to make sure that long-term cleanup activities continue and can be done promptly	8.0 (2.5)	45
ORG-3: Maintain role of citizen advisory boards that currently represent the community interests at the site	7.7 (2.5)	35
SR-3: Make sure remains owned by the federal government forever	7.6 (2.9)	45
ORG-4: Provide information about what is happening at site using web sites and visits by site personnel to schools and community groups	7.5 (2.7)	34
ORG-5: Require government to appoint and fund an independent watchdog group that includes local people and scientists to monitor site activities and report results to the community	7.5 (2.8)	38
SR-4: Do not allow any new nuclear-related activity on the site	6.8 (3.5)	42
ORG-6: Require that site managers live near the site	6.3 (3.1)	22
SR-5: Provide access to public via guided site visits	5.3 (3.1)	14

Source: Adapted from Greenberg 2007a

interesting to observe that numbers 4 and 5 were providing specialized training and equipment to local responders and an early warning system. Monitoring ecological systems and nearby residents has lesser priority. The public preferences mirror a logical surveillance protocol that emphasizes the likely places where exposure will be observed first, second calls for strengthening the response capacity of local first responders and then looks for evidence of chronic exposures.

The site restriction responses emphasized avoiding unnecessary exposures by not disposing of the land until the site is not dangerous and keeping unwanted recreational visitors away. There was much less support for the federal government ownership of the land forever and even less support for not allowing new nuclear-related activities on the site. The latter responses suggest a concern about limiting regional economic opportunities among some respondents.

The ideas for six organizational options came from conversations with citizen groups, and yet, these were the least favored. The most favored one of them was requiring the federal government to routinely report to site-specific advisory boards. Creating a trust fund and maintaining citizen advisory boards were favored by some respondents. Providing information via the web and by visits from site personnel, funding a watchdog group to support and inform the public, and requiring that site managers live near the facilities had some support but were among the least consistently favored. Indeed, some respondents considered them very low priorities.

4.5.3   Question 3. Correlates

4.5.3.1   Summary of Correlate Data

Table 4.5 provides data from our 2005 survey regarding 15 items to worry about the local DOE sites. The results did not change much through the year 2010 survey. Eight of the 15 showed differences between worried a "great deal" and worried "not at all" of 16 % or more. All eight are about possible human exposures from managing stored waste. Fear of a terrorist attack was an exception. The remaining seven also included one exception, which are explosions and fires resulting from cleanup. The remaining six less troublesome ones are about ecological, cultural, and economic impacts.

Table 4.6 lists feelings and emotions articulated by residents living near the six DOE sites in our 2010 survey. Reiterating, these were open-ended probes recorded by the interviewers and interpreted and classified by the first author into positive, negative, and "other" categories. Indeed, we had assumed that negative responses would outnumber positive ones. Feelings like safe and happy and economic ones related to jobs, development, and pride were frequently articulated. Fear, scared, concerned, and dangers were among the most common negative feelings and emotions, followed by lack of trust and suspicion (see also Table 4.8). There were 618 responses (29 %) of no feeling or emotional response. The most important

Table 4.5 Fifteen possible worries about DOE-related activities, 2005

Worry category ($n = 1,351$)	Worry a great deal, % (A)	Worry not at all, % (B)	Difference (A–B)
Materials at site will seep into the ground and contaminate local streams and drinking water	48	12	36
Disturbing corroding or leaking storage containers will cause more damage to the environment than leaving them in place	37	14	23
Workers involved in the cleanup will be exposed to hazards	37	14	23
There may be accidents when hazardous waste materials are transported to and from the site	36	14	22
Cleanup of chemicals or radioactive materials will expose residents to hazards	37	17	20
The site might be a target for a terrorist attack	39	19	20
Opening the site to public access could expose people to hazards	36	18	18
People may get sick from eating fish and wildlife from the site	37	21	16
Fish, bird, and animal habitats will be destroyed during the cleanup of the site	31	21	10
Some waste materials will need to remain buried or contained on the site	29	20	9
Animals such as deer and raccoons will carry contamination off the site	28	21	7
About new activities at the site that involve the use of nuclear materials	28	22	6
Area residents will lose jobs if the site reduces its cleanup efforts and other activities	24	18	6
High risk of explosions and fires related to the cleanup	25	26	-1
Some burial or archeological sites may be destroyed	19	30	-11

(1 = worry a great deal, 2 = some, 3 = not much, 3 = 4 = worry not at all)
Source: adapted from Table 2, Greenberg et al. (2007b)

observation is the large number of positive feelings and emotions (756, 36 %) compared to negative ones (546, 26 %).

It is especially interesting to compare the site-specific results to the national survey results for the same question. The national sample of 651 asked for feelings and emotions associated with a nearby energy, waste management, or manufacturing site. Many more national survey respondents offered no feeling or emotion (48.5 %) compared to the site-specific one (29.4 %). The proportions who stated a negative emotion was the same, 26 %, in both groups. The positive feelings were far more prevalent in the DOE site-specific survey (36 %) than in the national one (19 %).

The fact that positive emotions and feelings outnumbered negative ones does not necessarily mean that the positive images were stronger correlates of siting and site management options than negative feelings and emotions (see below).

Table 4.6 Feelings and emotions used to describe local department of energy site, 2010

Feelings and emotions ($n = 2,100$)	Number[a]	%
Positive		
Safe, good, happy, glad, great, positive	265	12.6
Jobs/employment, economy, pride, necessary, important to national security	264	12.6
Trust, reliable, comfortable, confident, dependable, secure	61	2.9
Accept it, no concern, no problem	54	2.6
Exciting, interesting, smart science, technology development	49	2.3
Attractive, remote, beautiful, big, peaceful, quiet, secluded, pleasant	49	2.3
Hopeful, optimistic, have faith	14	0.7
Positive total	756	36.0
Negative		
Fear, scared, worried, danger, anger, annoyed, bad, catastrophe, disaster, disgusted, furious, mad, pain, unsafe	314	15.0
Distrust	108	5.1
Concerned, nervous, suspicious	78	3.7
Hazardous, dirty, polluted, ugly	33	1.6
Disappointed, sad	13	0.6
Negative total	546	26.0
Other		
Neutral, apathetic, indifferent, emotionless, do not care	40	1.9
Not thought about it, none	36	1.7
Cautious	24	1.1
Secretive, mysterious	23	1.1
Ambivalent, confused, mixed emotions, perplexed	18	0.9
Curious	17	0.8
Historical, DOE, agency role	14	0.7
Calm	8	0.4
No feeling or emotions listed or not appropriate	618	29.4
Other total	798	38.0
Total	2,100	100.0

[a]Raw numbers add to 2100; % add to 100 %. Subtotals also add to 2100 and 100 %

We also asked respondents to tell us the first image and color that came to mind. Some of the pairings were notable. For example, there were 2.1 times as many pairings of fear with the color red than would have been expected by chance, twice as many pairings of fear and black than expected by chance, and 2.7 times as many nominations of fear and the image "polluted" than expected by chance.

Information about affect is more useful if it is viewed in the context of other environmental issues. Table 4.7 compares worrying about cleanup of nuclear waste and a nuclear power plant accident with natural disasters, traffic congestion, terrorism, polluted drinking water, loss of open space, and global warming. These represent both acute and chronic environmental issues. The two nuclear issues were ranked at the bottom of local environmental concerns by the national sample respondents. In comparison, DOE site respondents were more worried about the

Table 4.7 Public worries about local environmental issues, 2008, pre-Fukushima (% worried a great deal or some about issue)

Site	Nuclear power plant accident	Cleanup of nuclear waste	Natural disasters	Traffic congestion	Terrorism	Polluted drinking water	Loss of open space	Global warming
National sample (n = 600)	37[a]	37[a]	61[a]	65[a]	38	49	54	59[a]
Six DOE sites (n = 1,147)	47	52	45	52	40	49	51	52

"I'm going to read you a list of potential environmental issues in your area. For each please tell us how much each worries you, meaning does it worry you a 'great deal,' 'some,' 'not much,' or 'not at all.'"

[a]National sample value significantly different than site-specific one at $p < 0.05$

Table 4.8 Trust of key parties, 2008

Trust category (n = 2,101)	% Respondents strongly agree or agree
I trust community representatives (watchdogs) to inform us	82
I trust the competence of independently watchdog scientists who report to the community	78
I trust the federal agencies, such as the Nuclear Regulatory Commission, the Environmental Protection Agency, and the US DOE to competently oversee health and safety at the site	74[a]
I trust the federal agencies, such as the Nuclear Regulatory Commission, the Environmental Protection Agency, and the US DOE to competently communicate information to the community	72[a]
I trust my state government to communicate information to the community	68
I trust my state government to competently manage health and safety at the site	68
I trust local journalists to accurately tell us what is happening at the sites	65
I trust the private owners and operators to competently manage health and safety at the site	61[b]
I trust the private owners and operators to communicate information to the community	57[b]

1 = strongly agree, 2 = agree, 3 = neither agree or disagree, 4 = disagree, 5 = strongly disagree

[a]Federal agencies were trusted more than state, local journalists, and private owners and operators at $p < 0.05$

[b]Private owners and operators were significantly less trusted than others at $p < 0.05$

cleanup of nuclear waste and nuclear power plant accidents than their national counterparts and less about natural disasters, global warming, and traffic congestion ($p < 0.05$).

Losing trust can damage an organizations ability to manage. Table 4.8 shows how much critical actors were trusted. At the top of the list, 82 % and 78 %, respectively, trusted their community advisory board representatives and independent scientists. Next, federal agencies (DOE, NRC, and EPA) enjoyed considerable

Table 4.9 Cultural variable responses, 2009

Question ($n = 2,400$)	% Respondents strongly agree or agree
Too many people expect society to do things for them that they should be doing for themselves	90
I feel that people who are successful in business have a right to enjoy their wealth as they see fit	83
Discrimination against minorities is still a very serious problem in our country	73
The government interferes far too much in our everyday lives	69
Our society would be better off if the distribution of wealth was more equal	52
A lot of problems in society come from the decline in the traditional family, where the man works and the woman stays home	48

1 = strongly agree, 2 = agree, 3 = neither agree or disagree, 4 = disagree, 5 = strongly disagree

trust, 74 % and 72 %, respectively, to manage health and safety at the sites and to communicate to the public about these issues. State government was slightly less trusted, and journalists were trusted to tell the public what was happening by 65 % of respondents. At the bottom of the list were facility owners and operators, over 13 % below the federal agencies with regard to competently managing health and safety (61 %) and communicating information (57 %). The national agencies were significantly more trusted than the state agencies, the journalists, and the owners and operators ($p < 0.05$).

With regard to cultural and worldviews, we tested six of Kahan et al.'s variables as correlates. Table 4.9 shows that 90 % agreed that too many people expect society to do things for them that they should be doing for themselves, and 83 % felt that people who are successful in business should have the right to enjoy their wealth as they see fit. In contrast, only about half the respondents agreed that our society would be better off if the distribution of wealth was more equal and that a lot of problems in society come from the decline in the traditional American family. By themselves these numbers offer no information about their relationship with nuclear-related technologies.

Table 4.10 summarizes demographic characteristics using the unweighted data from the year 2005. We picked Rocky Flats and Savannah River to illustrate the variation among the DOE site regions. The Rocky Flats group is notably different from the Savannah River one with regard to education, race/ethnicity, and income. This shows why it was essential that weighting be done with regional numbers rather than with national ones.

The personal history and preference questions were too numerous to fully display. Instead, we present two illustrations that are directly relevant to the post-Fukushima survey in Chap. 5. Public preferences for nuclear waste management technologies depend on other parts of the nuclear fuel cycle. In 2010, for example, we asked if the USA should increase or decrease reliance on nuclear energy (stay the same was also accepted as an answer, but not often chosen). In the six DOE site regions, 52 % wanted to increase reliance and 44 % wanted to decrease reliance.

Table 4.10 Illustrative demographic characteristics of respondents, year 2005 survey

Demographic characteristics	All respondents DOE sites ($n = 1,351$)	Savannah River ($n = 225$)	Rocky Flats site ($n = 225$)
Age, years			
18–29	11.6	7.6	10.7
30–49	37.6	40.7	37.6
50–64	32.1	34.0	36.7
65+	18.7	17.7	15.7
Education			
College graduate	20.6	27.4	14.7
Race/ethnicity			
Asian	0.5	0.5	0.5
Black	8.0	27.4	5.0
Indian	2.7	1.8	3.6
Latino	3.6	1.8	8.1
White	85.4	68.9	83.3
Family income			
$1,000 s			
<25	19.9	18.3	16.8
25–49.9	30.4	32.2	24.5
50–74.9	23.2	25.5	22.1
75–99.9	12.5	10.6	18.8
100+	12.1	10.6	16.3
No answer	2.0	2.9	1.4
Sex			
Male	48.0	48.0	48.2

Source: Greenberg et al. 2007a

Table 4.11 Association between preference for reliance on nuclear energy and concerns about parts of the nuclear fuel cycle, 2010

Issue ($n = 2,400$)—US should increase reliance on nuclear power and	Contingency coefficient
Nuclear power plants could have a serious accident	0.361[a]
Nuclear waste is safely managed	0.401[a]
Nuclear facilities are vulnerable to a terrorist attack	0.316[a]
Nuclear material transporters dangerous for those living along the transport path	0.394[a]
Uranium mining degrades animals, plants, land, and water	0.359[a]

Source: Adapted from Greenberg and Truelove 2011
[a]Coefficient is significant at $p < 0.01$

Table 4.11 shows the relationship between these preferences for nuclear energy and parts of the nuclear life cycle (prior to the Fukushima events).

What is striking about the data is that the strongest predictor of preference for greater reliance on nuclear energy was risk beliefs about nuclear waste management. Only 39 % of those who did not think the waste was safely managed preferred increasing reliance on nuclear energy compared to over 80 % who thought it was safely managed. The second strongest association was with transportation. In 2010, prior to Fukushima, a serious accident at a power plant was only the third strongest correlate of preference for nuclear power. Even if this result is partly explained by the reality that the sample audience lives near nuclear near waste management sites, the information is noteworthy for DOE's site managers. The key message is that public preferences for new nuclear power plants are not independent of their views of the entire nuclear fuel cycle. The same reality is also true of coal, where the common belief that global warming is the key driver of public opinion about coal obscures the reality that many members of the public are extremely concerned about worker health and safety and ecological impacts (Greenberg and Truelove 2011).

4.5.3.2 Summary of Correlations with Preferences for Siting and Waste Management Policies

Over 100 binary logistic, ordinal, and ordinary least square regression analyses linked future site use for expanded nuclear missions, management of existing hazards, and the five sets of predictor variables. It was not possible to choose one or two to represent all of these. Instead, Tables 4.12 and 4.13 are a composite list of all the variables that were selected by a stepwise regression model at least once ($p < 0.05$ requirement for inclusion).

The tables paint a portrait of those that prefer new on-site activities and also see less need for stronger environmental and risk management policies. These respondents were much less worried than their counterparts about existing on-site activities and transport of nuclear materials and in general were optimistic about the environment in their area. Some viewed nuclear energy as clean energy, placed emphasis on jobs and the economy, and described the site with positive images. Disproportionately, these respondents were male, white, college educated, and affluent, and they trusted site managers to protect health and safety and communicate. They tended to describe themselves as political independents and relied on personal contacts and their own research rather than the mass media for information about their nearby site.

Their counterparts worried about multiple activities at the legacy sites and transport of nuclear materials, did not like nuclear energy, had negative images of their nearby site, and did not choose new jobs as a high priority. Whereas many proponents saw beauty, jobs, and open space, often based on personal experience, these respondents who were not familiar with the sites had images of hazards and ugly pollutants. Disproportionately, opponents of new missions were relatively poor African American and Latino women who tended to believe that discrimination remains a serious problem and that we would be better off if wealth were more equally distributed. These respondents were less trusting of site managers, especially contractors.

While the same correlates tended to be in both sets of regressions, there were several interesting differences between them. Feelings–emotions–worries

Table 4.12 Strongest correlates with preference for siting new nuclear facilities and increasing activities

Variable category	Variable
Emotions, feelings, images, colors	Not worried about local nuclear facility having an accident Low score on site worry scale
	Not afraid of facilities (open-ended)
	Facilities are safe (open-ended)
	Nuclear energy is clean energy (open-ended)
	Not worried about environmental problems in the area
	Facilities bring jobs and employment (open-ended)
	Image of nuclear site is natural environment, open space, and color brown (open-ended)
	Positive image of sciences and researchers (open-ended)
Trust	Trust competence of independent watchdog scientists to gather information and report it to local community groups
	High score on trust scale
	Trust federal government authorities to competently oversee health and safely Trust private operators and managers to competently oversee health and safely
Demographic	White respondent
	Male respondent
	High income respondent
	65+ years old
	College graduate
Cultural and values	Do not agree that discrimination against minorities is a very serious problem in the USA
Personal history and preferences	Respondent is familiar with nearby nuclear facility
	Respondent lives in a host county
	Economic impacts of nearby nuclear facilities are positive
	US should rely more on nuclear, solar, and wind power
	Coal is a harmful energy source
	Global climate change has had impact on view of nuclear power
	Local environment will be better in 25 years
	Identified as Republican or political independent
	Identifies as a resident of Idaho site region
	Respondent relies on personal contacts, books, articles, and web for information about nuclear technology, not on mass media

[a]Variables were significant correlates in at least one stepwise regression analysis with 2005, 2008, 2009, and 2010 data. Regressions were binary logistic regression or ordinal regressions with Nagelkerke R2 estimates ranging from 0.116 to 0.222
Open-ended = open-ended question

dominated the legacy wastes management regressions. Seventeen of the 31 variables in Table 4.13 were from the worry list. The public wanted actions to deal with their very specific concerns captured in the long lists of worries about the legacy wastes. Preferences about future on-site activities also had a strong presence of feelings and emotions (9 of 29), but many of these are positive and intersect with personal history and related preferences.

Table 4.13 Strongest correlates with preference for not having stronger environmental and risk management policies[a]

Variable category	Variable
Emotions, feelings, images, colors	Not worried about local nuclear facility having an accident
	Low score on site worry scale
	Not afraid of facilities (open-ended)
	Facilities are safe (open-ended)
	Nuclear energy is clean energy (open-ended)
	Not worried about environmental problems in the area
	Not worried about disposal of toxic and mining wastes
	Not worried about a terrorist attack at the site
	Not worried that disturbing waste will cause major exposures
	Not worried that opening the site to public could lead to public exposures
	Not worried about transportation accidents involving hazardous waste materials
	Not worried about cleanup worker exposure
	Not worried about fire and explosions related to site cleanup
	Not worried about destruction of buried or archeological sites
	Respondent not worried or concerned about site
	Site image is jobs and employment (open-ended)
	Site image is natural environment (open-ended)
Trust	High score on trust scale
	Trust federal government authorities to competently oversee health and safety at the site
	Trust federal government to competently communicate information to the community
	Trust was an emotional reaction to site (open-ended)
	Local government does protect public in the area
Demographic	Male respondent
	High income respondent
	College graduate
Cultural and values	Disagrees that society would be better off if wealth were more equally distributed
Personal history and preferences	Respondent is familiar with nearby nuclear facility
	Respondent lives in a host county
	Respondent is optimistic about the local environment
	Identifies as a resident of Idaho site region
	Respondent relies on personal contacts, books, articles, and web for information about nuclear technology, not on mass media

[a]Variables were significant correlates in at least one stepwise regression analysis with 2005, 2008, 2009, and 2010 data. Regressions were ordinary least squares with R^2 estimates ranging from 0.326 to 0.684

The second point of additional emphasis is an age-related finding that is not apparent in these aggregate results but is important because it ties back to the role of waste management in the nuclear fuel cycle. American senior citizens were alive during the administrations of Franklin Roosevelt, Harry Truman, and Dwight

Eisenhower. Those were years when new coal, oil, and gas findings were frequent and nuclear power was touted as an inexpensive and almost limitless source of energy. The challenge of managing these wastes was not widely understood nor publicized. We found relatively strong support among senior respondents for nuclear power and coal, and relatively less support for wind, solar, and other "new" renewable sources of energy. For example, our 2008 survey found that 40, 63, and 36 % of those 65+ years old favored greater reliance on coal, nuclear, and oil as a fuel compared to 27 %, 33 %, and 15 % among those who were <35 years old.

During the 1940s and extending well into the 1960s, the US built health care and social programs focused around what today would be considered liberal elements of the Democratic and Republican parties. Our reason for noting this finding is that today's younger populations do not necessarily share these priorities and views of the role of government in stimulating the economy as well as supporting new energy technologies. Within 20 years, this younger cohort will be politically and demographically dominant. What support they offer to develop nuclear energy, new nuclear science, and allocating resources to manage the nuclear waste legacy is an open question.

4.6   Summary and Lessons Learned

A final context for the 2011 survey in Chap. 5, this chapter briefly summarizes the limitations and results of the surveys we conducted in 2005, 2008, 2009, and 2010.

4.6.1   Limitations

L1. RDD surveys during the last decade increasingly faced much lower response and cooperation rates, requiring more expensive callback designs and weighting protocols. Now they require a cell phone component.

L2. Four of six DOE-centered regions were in every survey and five others were in some but not others.

L3. Sample sizes for individual surveys ranged from 190 to 350, which implies relatively high standard sampling error for each site result but not for the national sample or the aggregate site-specific results.

L4. Survey length was 18–22 min, which limits the numbers of questions but means that responses should have minimal respondent fatigue.

4.6.2   Results: New Activities at Legacy Sites

R1. Two-thirds of site-specific sample favors DOE's energy park concept, with 36 % favoring it for their own area.

R2. Site-specific support for new waste management activities was >50 % and was higher than national sample. Proportions favoring new laboratory and nuclear energy sites were almost identical to the national sample.

R3. The Idaho site, host county residents, and those who are familiar with the site because they worked there or had a relative who did were the strongest supporters.

R4. Fernald, Mound, and Rocky Flats were closing or were closed, and their populations were the least supportive of new on-site activities.

4.6.3   Results: Legacy Waste Management

R5. Respondents preferred environmental and risk-based management policies that monitor the water and air at the site, workers, and strongly supported equipment and training for local first responders and tools to alert the public.

R6. Requiring site managers to live near sites, prohibiting new missions, providing guided tours of the sites, and other organizational steps were the least favored priorities.

R7. Legacy site environmental and risk management preferences were stable between 2005 and 2010.

4.6.4   Results: Correlates

R8. Emotions and feelings were the strongest correlates, especially with regard to environmental management options. The site-specific sample had many more positive feelings, emotions, and images than the national sample and many more links to positive economic outcomes.

R9. Trust was relatively high, averaging about 70 % for federal agencies like the DOE, NRC, and EPA and was 13 % or more lower for contractors and owner-operators.

R10. Two demographic patterns stood out. One was so-called white male effect with affluent college-educated white males disproportionately supporting new on-site activities and the second a group of relatively poor minority respondents who were less supportive and wanted very high levels of environmental and risk management for the legacy sites.

R11. Other factors, most notably personal familiarity, use of personal contacts, the web and non-media sources, a history of having lived through the period of great promise for nuclear and fossil fuel energy, and optimism about the future were also associated with strong support for new energy-related activities at the legacy sites.

4.6.5   Policy-Related Results

R12. Prior to Fukushima, there was considerable local public support for new missions and intensification of existing ones.

R13. The strongest supporters typically were politically powerful and knowledgeable white males who live close to site and benefit from their presence.

R14. Positive feelings and images among site respondents were disproportionately positive compared to other studies and the national sample.

R15. Prior to Fukushima, many respondents were becoming more willing to consider more reliance on nuclear technology because it is perceived as a lesser evil than fossil fuels.

R16. Respondents had high expectations of the federal agencies but also had a high level of trust of these agencies to both protect them and communicate with them.

R17. There are multiple publics with different concerns and backgrounds that need to be sought out in order to build and maintain a cooperative environment with the individuals who will be living with the DOE legacy sites for many generations.

References

Cantor J, Brownlee S, Zukin C, Boyle J (2009) Implications of the growing use of wireless telephones for health care opinion polls. Health Serv Res 44(5):1762–1772

CTIA, the Wireless Association (2011) Wireless Quick Facts. http://www.ctia.org/advocacy/research/index.cfm/aid/10323. Accessed August 5, 2011

Curtin R, Presser S, Singer E (2000) The effects of response rate changes on the index of consumer sentiment. Public Opin Q 64:413–428

Greenberg M (2009a) NIMBY, CLAMP and the location of new nuclear-related facilities: U.S. National and eleven site-specific surveys. Risk Anal 29(9):1242–1254

Greenberg M (2009b) Energy sources, public policy, and public preferences: analysis of US national and site-specific data. Energy Policy 37:3242–3249

Greenberg M (2009c) What environmental issues do people who live near major nuclear facilities worry about? Analysis of national and site-specific data. Environ Plann Manage 52(7):919–937

Greenberg M (2010) Energy parks for former nuclear weapons sites? Public preferences at six regional locations and the United States as a whole. Energy Policy 38:5098–5107

Greenberg M, Lowrie K (2002) External stakeholders' influence on the DOE's long-term stewardship programs. Fed Facil Environ J, Spring; 65–75

Greenberg M, Truelove H (2010) Right answers and right-wrong answers: sources of information influencing knowledge of nuclear-related information. Socioecon Plann Sci 44:130–140

Greenberg M, Truelove H (2011) Energy choices and perceived risks: is it just global warming and fear of a nuclear power plant accident? Risk Anal 31(5):819–831

Greenberg M, Lowrie K, Burger J, Powers C, Gochfeld M, Mayer H (2007a) The ultimate LULU? Public reaction new nuclear activities at major weapons sites. J Am Plann Assoc 73 (3):346–351

Greenberg M, Lowrie K, Burger J, Powers C, Gochfeld M, Mayer H (2007b) Nuclear waste and public worries: public perceptions of the United States major nuclear weapons legacy sites. Hum Ecol Rev 14(1):1–12

Greenberg M, Lowrie K, Burger J, Powers C, Gochfeld M, Mayer H (2007c) Preferences for alternative risk management policies at the United States major nuclear weapons legacy sites. J Environ Plann Manage 50(2):187–209

Greenberg M, Lowrie K, Hollander J, Burger J, Powers C, Gochfeld M (2008) Citizen board issues and local newspaper coverage of risk, remediation and environmental management: Six United States nuclear weapons facilities. Remediation 18(3):79–90

Greenberg M, Mayer H, Powers C (2011) Public preferences for environmental management practices at DOE's nuclear waste sites. Remediation 21(2):117–131

John F. Kennedy School of Government (2009) Civic Engagement in America. http://www.cfsv. org/communitysurvey/results10.html, last accessed May 28, 2009

Keeter S, Miller C, Kohut A, Groves R (2000) Consequences of reducing nonresponse in a national telephone survey. Public Opin Q 64:125–148

Lowrie K, Greenberg M (2000) Local impacts of US nuclear weapons facilities: a survey of planners. Environmentalist 20(2):157–168

Lowrie K, Greenberg M (2001) Can David and Goliath get along?: Federal lands in local places. Environ Manage 28(6):703–711

Lowrie K, Greenberg M, Waishwell L (2000) Hazards, risk, and the press: a comparative analysis of newspaper coverage of nuclear and chemical weapons sites. Risk: Health, Safety Environ 49:49–67

Merkle D, Edelman M (2002) Nonresponse in exit polls: a comprehensive analysis. In: Groves RM, Dillman DA, Eltinge JL, Little RJA (eds) Survey nonresponse. Wiley, New York, pp 243–257

Pew Research Center (2004) Survey report/press release: polls face growing resistance, but still representative. Pew Research Center for the People and the Press, Washington, D.C

Public Opinion Quarterly (2006) Nonresponse bias in household surveys. Special Issue 70(5)

The American Association for Public Opinion Research (2008) http://www.aapor.org/ responseratesanoverview, last accessed June 10, 2009

The American Association for Public Opinion Research (2009) Standard definitions: final dispositions of case codes and outcome rates for surveys, 6th edn. AAPOR, Lenexa, Kansas

Zukin C (2006) The future is here! Where are we now? And how do we get there? Public Opin Q 70(3):426–442

Chapter 5
Impact of the Fukushima Events on Public Preferences and Perceptions in the United States, 2011

Abstract The CRESP 2011 post-Fukushima survey included a cell phone component of 25 % in order to avoid potential bias associated with landline only surveys. Sample sizes for the six site-specific studies (Hanford, Idaho, Los Alamos, Oak Ridge, Savannah, WIPP) were 180 (a total of 1,080 respondents), and the national sample size was 850.

Preferences for new nuclear-related activities in respondents' host states dropped from 48 % in 2010 to 33 % in 2011. Everyone of the sites showed a substantial decrease. Almost 60 % of national respondents and 73 % site-specific ones favored nuclear power. However, two-thirds of these are more worried about nuclear power than before the events in Japan (see below). Between 2010 and 2011, the proportions favoring greater reliance on nuclear power for electrical energy decreased approximately 15 %, whereas the proportion favoring natural gas increased almost 10 %. Concerns about safety and exposures, and associated emotions and feelings were the strongest correlates of support for nuclear technology.

In 2010, trust of DOE and other responsible parties averaged about 70 % for federal agencies and contractors. One year later, after Fukushima, trust fell approximately 10 %. Trust was the strongest or second strongest correlate of many of the results, a stronger role than in the earlier surveys.

A new important dynamic in the 2011 data is the appearance of a large group (43 % of respondents) that supports nuclear energy and new nuclear-related facilities but has become more equivocal in their support after Fukushima. This group disproportionately is concerned about global climate change and trusts DOE, but that trust is lukewarm. This group is an important one to address if the DOE, NRC, and other pro-nuclear industry groups want to maintain support for nuclear energy. Without continuing support from this group, nuclear energy would have preference levels similar to coal and oil.

M.R. Greenberg, *Nuclear Waste Management, Nuclear Power and Energy Choices,*   93
Lecture Notes in Energy 2, DOI 10.1007/978-1-4471-4231-7_5,
© Springer-Verlag London 2013

5.1   Introduction

The March 2011 earthquake, tsunami, and damage to multiple Japanese commercial nuclear reactors have caused governments to reevaluate nuclear power. How has the US public responded? Chapter 3 reviewed pre- and post-Fukushima public surveys, and Chap. 4 summarized results from the CRESP 2005, 2008, 2009, and 2010 pre-Fukushima surveys. This chapter presents the results of the CRESP 2011 post-Fukushima survey, which measured the US public reaction to the Fukushima events with emphasis on the DOE sites.

The survey focused on two questions:

1. After Fukushima, has there been a noticeable change in public preferences and perceptions about nuclear-generated energy and waste management practices in the USA? This is the preferences and perceptions question.
2. Are these changes associated with environmental concerns, trust, cultural and worldviews, demographic characteristics, and personal history and preferences of respondents? This is the correlates question.

5.2   Design of the Survey and Questions

5.2.1   Questions

The survey was designed around two blocks of questions. One measured changes in public preferences and perceptions about nuclear energy and nuclear waste management, and examined specific policy issues related to siting new nuclear activities and handling spent fuel. The second block included questions to measure the five sets of correlates presented in Chaps. 3 and 4: emotions/worries/concerns, feelings, trust, cultural and worldviews, demographic attributes, and respondent personal history and preferences.

5.2.1.1   Preferences and Perceptions

The first three questions asked about the impact of the events in Japan on perception of nuclear power. All of these responses were randomized by the caller to avoid order bias.

"Nuclear energy generates about 20 % of the United States electrical energy use. Keeping that in mind, please tell me how the nuclear power plant events in Japan earlier this year influenced your opinion on nuclear energy production here in the United States. I am going to read you four statements and tell me, please, which one statement most closely fits your viewpoint.

1. I remain a firm supporter of increasing nuclear power.
2. I am still a supporter but I am more concerned than before the events in Japan
3. I was open-minded about nuclear power, but am now against it.
4. I have been opposed to nuclear power and these events have not changed that opinion."

In 2010 and 2011, we asked "how much has your opinion on nuclear energy been influenced by your concern about global climate change."

1. I have been a firm supporter of increasing nuclear power.
2. I have opposed nuclear power and my concern about global climate change has not changed that opinion.
3. Global climate change has made me more open to considering new nuclear energy facilities.

In multiple surveys, we asked respondents' preferences for different electrical energy sources.

"For each of the following forms of electric power generation, please tell me whether you think the United States should increase or decrease our reliance on it." Stay the same was permitted as a voluntary response. The options were coal, dams or hydro, natural gas, nuclear, oil, solar, wind, and biomass or biofuels.

The next set of questions focused on use of the DOE's major waste management sites. The first is the new on-site energy activities question, which asked about siting of new energy, waste management, and science facilities.

1. I favor it and am willing to have it in my state
2. I favor it but want it in another state
3. I favor it but have no preference about where the site would be located
4. I am neutral to the idea
5. I am against the idea

Before the Japanese events, few Americans knew anything about so-called commercial "spent" fuel. Asking about this subject required separating the issues into comprehensible parts.

When nuclear fuel is no longer useful for energy production, it is still radioactive and has to be treated very carefully. It is moved to large tanks filled with cooling water called used fuel pools. What happens next is up to the particular nuclear plant. Management decides how long it stays in the storage pools and how long to wait before transferring it to concrete casks for long-term storage.

Our first question probed spent fuel as an issue that many members of the public only learned about because of the events in Japan. Should "the United States ... maintain the current policy of letting each separate nuclear facility decide when to transfer used fuel from the storage pools to in concrete casks," or "the United States should require the transfer of all used fuel from the storage pools to concrete casks as soon as practically achievable? So, what's your opinion on that? Should the United States

1. Maintain the current policy of letting each separate nuclear facility decide when to transfer used fuel from the storage pools to in concrete casks

2. Require the transfer of all used fuel from the storage pools to concrete casks as soon as practically achievable."

Then we asked about long-term storage options.

"We have four alternative options the long-term storage of used nuclear fuel in concrete casks. I am going to read those options to you, and please tell me which one you prefer.

1. Transfer all used fuel in casks to three or four locations in the USA that already manage and have the US defense-related nuclear waste for temporary storage.
2. Transfer all used fuel in casks to three or four locations in the USA that would be developed as used fuel storage facilities for temporary storage.
3. Move all used fuel in concrete casks to a single deep underground repository at Yucca Mountain, Nevada, that was partly built but not completed.
4. Move all used fuel in concrete casks to a single deep underground repository location that has not yet been identified."

Transportation is a major public concern. Hence, we asked.

"If the nuclear industry needs to move used commercial nuclear fuel across the United States, how should they do it? Should they do it by truck over highways, by railroad over rails, or by barge over waterways. So, please tell me, of these three, which is your most preferred?

1. By truck over interstate highways
2. By railroad or rail
3. By barge over waterways"

And then we asked which was their least preferred? We examined the answers to each of these questions separately.

5.2.1.2   Correlates

The correlates were grouped into five sets as in Chaps. 3 and 4. The same grouping was maintained in 2011. Some changes to individual questions were made and those are described here.

Emotions, feelings, images and colors were expected to be the strongest correlates of preferences and perceptions. We asked about knowledge and concern about commercial nuclear power plants and the US Department of Energy waste management sites. First we asked if they knew if there was a US Department of Energy nuclear waste site in their state. If they answered yes, we asked how worried they were about those in their state. If they did not know, then we asked about nuclear waste sites, in general. Note that the vast majority of respondents have had one in their state.

"Please tell me about your level of concern about [name the site]. On a scale from 1 to 10, where 1 is 'not at all worried,' and 10 is 'extremely worried,' how much do you worry about [name the site]."

Then, the exact same question was asked about a nuclear power plant in their state. These questions were a quantitative base for the analysis.

The second set of questions was a modification of the open-ended questions about feelings, emotions, images, and colors in the 2010 survey. The objective was to ask how memorable Fukushima was compared with other major hazard events.

"We are interested in how well you recall certain events. I'm going to read a list of six events each of which occurred within the last 10 years. For each event, please tell me how memorable it was to you. By memorable, I mean sitting here and talking now, how much can you remember about the event. The choices are that you 'don't remember anything,' that you 'remember a few details,' that you 'remember many details,' or that you 'remember many details' and you 'clearly remember' where you were when it happened." The list of six events was randomized.

(a) The World Trade Center attack in New York City in 2001
(b) Hurricane Katrina in New Orleans in 2005
(c) The offshore oil drilling platform blowout and oil leak in the Gulf of Mexico in 2010
(d) The coal ash spill in Eastern Tennessee in 2008
(e) The tsunami in the Indian Ocean that hit Thailand, Sri Lanka, Indonesia, and India in 2004
(f) The earthquake, tsunami, and nuclear power plant events in Japan in 2011

This list of six allowed us to compare Fukushima's memorability with others. The expectation was that Fukushima would be one of the memorable events along with the World Trade Center and Katrina and that the negative emotions would be associated with a high worry score. However, a high worry score might be independent of the emotions related to a specific event and not specially associated with a DOE waste management or commercial power plant site.

Trust was assumed to be a strong correlate. The same six questions were used in 2010 and 2011, and this becomes a pivotal set of questions because irrespective of respondent answers to the change questions, if trust in DOE has dramatically declined, then there is a problem in short-term and perhaps even long-term management of the accumulated waste.

Respondent demographic attributes were assumed to be strong predictors. As in our other surveys, respondents were asked about their age, educational achievement, ethnicity/race, and sex (see Chap. 3).

Five of the six statements of *culture, worldviews, and value* were used in 2011 and 2010 (see Chaps. 3 and 4).

Personal history was the fifth and last set of correlates. As in previous surveys, we asked about political party preference, concern about their local environment now and in the future, and familiarity with a local nuclear site (waste management and power plant). In addition, a number of new probes were asked in the 2011 survey to ascertain if personal experience and preparation were related to preferences and preferences. The first asked about personal experience with disasters.

"We'd like to know if you've ever had direct personal experience with a disaster. Have you personally ever experienced, by which I mean have you ever been in a hurricane, flood, earthquake, tornado, explosion, train derailment, mud slide, or major fire?" Then, "have you ever been evacuated from your home or work as a result of a disaster?" The answers were sorted into yes, no, and no, but have gotten warnings.

We also were interested in the importance respondents' attached to information and to economic issues associated with the nuclear industry.

"Let's say you wanted to compare the pros and cons of different energy sources. To do that, you'd need information about those energy sources. Of the different types of information available about energy sources, we'd like to know which information is most important to you. I'm going to read you six types of information about energy production and for each, please tell me whether that type of information is very important, somewhat important, or not important at all to you in comparing the pros and cons of different energy sources.

(a) Information about overall safety and accident history
(b) Information about the cost of energy production
(c) Information about the environmental impacts of building and operating energy production facilities
(d) Information about the degree of United States reliance on the energy type
(e) Information about the cost of the energy product to the consumer, for example, a gallon of gas or an electricity bill
(f) Information about waste, meaning what's left over after the energy production process"

We wanted to separate preferences based on what respondents believe society should use as decision-making criteria from their personal focus on economic issues related to energy: "How much does the price of gasoline, natural gas, and electricity influence your choices of the kinds of energy we should rely on?" The scale was a great deal; somewhat; not much, and not at all.

The last of the personal history and preference questions further probed the factors that influence US energy policy.

"For each of the following factors, please tell how much you think that policymakers should rely on that factor. In other words, please tell me whether in making energy policy United States policymakers should rely on each of the following factors 'a great deal,' 'somewhat,' 'not much,' or 'not at all.'

(a) Price to consumer
(b) Safety and accident history
(c) Energy independence
(d) Climate change
(e) Waste, meaning what's left over after the energy production process
(f) Environmental considerations
(g) Human health effects
(h) National security
(i) Economic considerations."

The practice of randomizing lists of choices was continued when administering these questions.

5.2.2   Sampling Locations

The regions chosen for the site-specific samples were Hanford, Idaho, Los Alamos, Oak Ridge, Savannah River, and WIPP. And a national sample was also collected (see Chap. 2 for more detail, maps, and photos of these sites).

5.2.3   Survey Implementation

Chapter 4 described and illustrated the sampling protocol used for 2005, 2008, 2009, and 2010. The 2011 protocol was identical, with one important difference, which was to include a cell phone sample. After summarizing the protocol, some space is devoted to explain the cell phone sample.

The survey questions were pretested on Wednesday and Thursday, June 22 and 23, 2011. After minor changes, the data gathering began on Wednesday, July 6, 2011, and ended on Friday, September 9, 2011. To avoid a seasonal sampling bias, these dates were deliberately chosen to be similar to those for the other CRESP surveys. The mean interview duration was 23.3 min for landline and 23.6 min for cell phone.

The site-specific samples included people who lived within a 50-mile radius of six designated nuclear sites that we had previously sampled: (1) Idaho National Laboratory (Idaho), (2) Oak Ridge National Laboratory (Tennessee), (3) Waste Isolation Pilot Plant (New Mexico, a/k/a "WIPP"), (4) Savannah River National Laboratory, (5) Los Alamos National Laboratory (New Mexico), and (6) Hanford site (Washington).

A ten callback design was used to achieve a desired response rate of 20 % and cooperation rate of 30 %. The combined response rate was 20.1 % (20.4 % for landline and 19.5 % for cell) and the cooperation rate was 29.9 % (30.2 % for landline and 29.4 % for cell). The total sample was 1,930 (1,445 landline and 485 cell phone).

This was the first of the CRESP surveys to include a cell phone sample. At the time this survey was designed, 25 % of the US residents only had cell phones. Hence, we designed this survey for a 75 % landline and 25 % cell sample. As noted in Chap. 4, the proportion is rapidly increasing. If we were repeating the survey in 2012, the split would be 70–30, and even that split underestimates the cell component because almost half of adult Americans only receive their calls on cell phones (Blumberg and Luke 2010; CTIA 2011).

To properly integrate the landline sample with the cell phone sample, the team used an overlapping dual frame RDD design under which a sample of telephone numbers was drawn from the landline sampling frame and a companion sample was drawn from a cellular sampling frame. The zip code is the geographical sampling frame for the landline survey, and the rate or billing center is the site for the cell sample. The survey was administered in both English and Spanish, with 2.4 % in Spanish, which is almost identical to those from previous CRESP surveys.

Sampling error is the likely difference in response between the sample and actual population for a chosen degree of statistical confidence. To be consistent with standard practice, we used a 95 % confidence interval. For this survey of 1,930, the overall sampling error was 2.2 %, and it was 3.4 % for the national sample of 850 and 3.0 % for the site-specific sample of 1,080. The six sites each had sample of 180, which means a sampling error of 7.3 %. In other words, we can be much more confident of the aggregate national and site-specific results than of each site-specific set.

As with other CRESP surveys, the results were weighted by age (18–44, 45–74, and 75+), sex, and race/ethnicity separately for each site-specific area and the nation as a whole.

The 2011 sample required joining the cell and landline surveys. Age differences were expected because cell users are disproportionately younger. Age differences per se are not a concern because the data can be merged. The concern is that cell and landline users, after controlling for age, provide different answers to the same questions. For example, it might be that 35–44-year-old cell phone user answers to questions about nuclear waste management are significantly different than those from 35- to 44-year-old landline respondents.

It is important that this possibility be checked to guard against the sampling method unduly influencing the results. Checking for this possibility requires the analyst to determine if the age-specific results are the same for the key indicators. If the answer is that the results are influenced by the sampling method, then the analyst must determine if the confounding is associated with one or more other factors that likely explain the result. For example, let us assume that the 75+-year-old population sampled via cell phone had a much stronger positive preference for a new nuclear power plant in their state than their landline counterparts. Further research might show that this finding is associated with greater socioeconomic status or another indicator. We found no statistically significant differences in the cell versus landline results. Hence, the age-specific results were combined.

5.3   Results

5.3.1   Question 1. Preference Responses

Table 5.1 measures the extent to which the Fukushima events changed public preferences for nuclear power. The table shows that 59 % of national level respondents favored nuclear power, even after the events. Notably, this compares to 57 % in a Gallup poll (see Chap. 3). What is particularly striking is that 70 % of supporters (national scale) said that they were still supportive, but were more concerned than before the events in Japan.

Table 5.1 Fukushima events and public preference for nuclear power, 2011

	Choice	Year 2011, %
Site-specific DOE sample (n = 1,046)	Remain firm supporter increasing nuclear power	27.2
	I am still a supportive but I'm more concerned than before the events in Japan	45.4
	I was open-minded about nuclear power but am now against it	14.8
	Have been opposed to nuclear power	12.6
National sample (n = 815)	Remain firm supporter increasing nuclear power	17.7
	I am still a supportive but I'm more concerned than before the events in Japan	41.5
	I was open-minded about nuclear power but am now against it	19.1
	Have been opposed to nuclear power	21.7

DOE site results show similar equivocation after the events in Japan, but they begin with a higher level of support for nuclear power—73 % site-specific support for nuclear power compared to 59 % national support.

Table 5.2 adds the dimension of global climate change. In 2010 and 2011, we measured the proportion of the population that had become more open-minded about nuclear power because of what they have learned about global climate change. The national and site-specific samples in both years showed that 38 % of respondents had become more open to considering nuclear power because of global climate change. However, the categories of supporter of nuclear power and opponents of nuclear power changed between 2010 and 2011, with supporters declining about 11 % and opponents increasing by about 10 %. In short, the Fukushima events did not eliminate the public's perceived benefit of nuclear power as one solution to global climate change but nuclear power support did decline.

Table 5.3 compares proportions in favor of increasing reliance on nuclear fuel and seven other options as an electrical energy sources. Several numbers stand out. At the national level, proportions favoring additional reliance on nuclear energy fell from 53 % in 2010 to 37 % in 2011. Notably, Pew polls in 2008, 2010, and 2011 show a fairly similar pattern at the national scale (see Chap. 3). The CRESP survey results in 2008, 2010, and 2011 to almost exactly the same questions showed 42, 53, and 37 % in favor of increasing reliance on nuclear power compared to 44, 52, and 39 % in the Pew (2011a, b) polls. In other words, there was an increase between 2008 and 2010 in public preference for greater reliance on nuclear power and that trend was stopped and reversed after the Fukushima events occurred.

The proportions supporting other electrical energy sources are at least as interesting as the ones for nuclear power. Responses for coal, hydroelectric, solar, and wind remained about the same. Preferences for natural gas jumped markedly from about half to about 70 % of respondents, and preferences for oil also increased. Practically speaking, natural gas appears to be the public choice for the baseline

Table 5.2 Opinion about nuclear power and global climate change, 2010 and 2011

Site	Choice	Year 2010, %	Year 2011, %
Site-specific DOE sample ($n = 1,956$ in 2010 and 1,317 in 2011)	Supporter of nuclear power	41.4[a]	30.4
	Opposed to nuclear power	21.2	31.4
	Global climate change made me more open to considering nuclear power	37.5	38.2
National sample ($n = 605$ in 2010 and 478 in 2011)	Supporter of nuclear power	33.4[a]	21.7
	Opposed to nuclear power	30.1	40.1
	Global climate change made me more open to considering nuclear power	36.5	38.2

[a]Site-specific aggregate sample favoring nuclear significantly higher than national sample at $P < 0.05$ but proportion more open because of global climate change not significantly different

Table 5.3 Preference for increasing reliance on as an electrical energy source, 2008, 2010, and 2011

Site	Choice	Year 2008, %	Year 2010, %	Year 2011, %
Site-specific DOE sample ($n = 1,062$ in 2008, 1,981 in 2010, 1,029 in 2011)	Coal	38.5	29.5	32.5
	Biofuels	NA	NA	68.4
	Dams/hydro	66.8	NA	69.7
	Natural gas	51.7	62.7	70.2
	Nuclear	49.0	62.7	48.7
	Oil	23.5	25.9	35.8
	Solar	92.7	NA	90.2
	Wind	89.9	NA	86.0
National sample ($n = 563$ in 2008, 610 in 2010, 794 in 2011)	Coal	33.2	34.9	34.9
	Biofuels	NA	NA	70.7
	Dams/hydro	73.1	NA	74.4
	Natural gas	52.0	65.9	67.7
	Nuclear	42.3	53.3	37.3
	Oil	21.5	21.0	26.3
	Solar	93.1	NA	93.6
	Wind	89.9	NA	89.2

source of electrical energy. How long that remains will be interesting to follow because of the controversy over fracturing and injecting liquids to obtain gas.

The Ipsos poll (2011) shows much greater worldwide support for hydroelectric, natural gas, and coal than in the USA. Perhaps, these US 2011 data signal that the US public is more amenable at least to natural gas than they have been in the past.

CRESP asked about public preferences for new nuclear-related activities at DOE sites in multiple surveys (see Chap. 4). The same wording was used in 2010 and 2011. Table 5.4 compares the year 2010 and 2011 results by site. The general pattern is for a substantial decline in support for new activities. Reiterating, these activities include such diverse functions as developing new technologies for solar,

Table 5.4 Preference for new energy-related activities at DOE sites, 2010 and 2011

Site	Choice	DOE sample, 2010, %	DOE sample, 2011, %
Site-specific DOE sample ($n = 2,060$ in 2010, 1,347 in 2011)	Favor in own state	47.9	32.6
	Favor in another state	10.2	8.6
	Favor with no location preference	20.3	21.5
	Neutral	16.5	27.5
	Against	5.1	9.8
Hanford ($n = 345$ in 2010, 175 in 2011)	Favor in own state	50.4	33.1
	Favor in another state	9.6	3.4
	Favor with no location preference	17.7	22.9
	Neutral	15.1	28.0
	Against	7.2	12.6
Idaho ($n = 348$ in 2010, 175 in 2011)	Favor in own state	66.8[a]	42.9
	Favor in another state	4.6	5.1
	Favor with no location preference	14.6	26.3
	Neutral	11.2	23.4
	Against	2.9	2.3
Los Alamos ($n = 336$ in 2010, 171 in 2011)	Favor in own state	44.8	31.6
	Favor in another state	7.2	11.1
	Favor with no location preference	20.6	13.5
	Neutral	21.5	28.1
	Against	6.0	15.8
Oak Ridge ($n = 345$ in 2010, 172 in 2011)	Favor in own state	47.8	39.5
	Favor in another state	11.3	5.8
	Favor with no location preference	23.5	18.0
	Neutral	14.8	22.7
	Against	2.6	14.0
Savannah River ($n = 342$ in 2010, 171 in 2011)	Favor in own state	38.8	21.1
	Favor in another state	13.7	17.5
	Favor with no location preference	23.9	21.6
	Neutral	18.1	35.7
	Against	5.5	4.1
WIPP ($n = 341$ in 2010, 176 in 2011)	Favor in own state	38.3	27.3
	Favor in another state	14.6	8.5
	Favor with no location preference	21.9	26.1
	Neutral	18.7	27.3
	Against	6.4	10.8

[a]Idaho support is significantly stronger than all other counterparts at $p < 0.05$

wind, and fuel cells to building new electrical generating capacity. While there is a doubling of opposition to the idea of new activities at the six DOE sites from 5 % to 10 %, the bigger shift is from "favor in own state" to "neutral" to the idea. This finding is consistent with the equivocation shown in response to nuclear power in Table 5.1 and in response to the role of global warming in Table 5.2.

From the Idaho site to the less enamored WIPP and Savannah River ones, every site, with the exception of Oak Ridge, had a reduction of 10 % points favoring new facilities in their own state. As part of the second research question, this chapter will examine the relationship between this result and trust, as well as other predictors (see below).

In 2008, we asked 2,701 US residents questions to test their knowledge of nuclear waste and power. One of the questions asked was what happens to nuclear fuel from power plants after it is no longer viable in reactors. Fifty-nine percent of those who lived near a major DOE site had no answer, which compared to 65 % of those who lived near a nuclear power plant and 71 % who lived near neither. The modal answer to our question was that spent commercial nuclear waste from reactors was stored underground at Yucca Mountain! Less than 10 % of respondents in 2008 replied that it was stored at commercial nuclear power plants.

With this historical information as context and with the issues associated with spent nuclear fuel brought to the public's attention as a result Fukushima, we asked respondents two questions about use commercial nuclear fuel. First, we asked if the USA should require the transfer of all used nuclear fuel storage pools to concrete casks as soon as practically achievable. Then, assuming that the USA desires to have some geographically centralized management of these wastes, we gave respondents four options for relocating this waste.

Tables 5.5 and 5.6 summarize the results. Almost 60 % prefer transferring all used commercial fuel to some sort of concrete casks as soon as practically achievable. Setting aside nonresponses, 67 % favor this policy. We have no data to directly compare this result to pre-Fukushima times. However, the relationship between this result and the Fukushima events is examined below; specifically we explore the relationship between the desire to transfer the waste to other locations and respondent characteristics.

Among those who favored transferring the casks to other locations in the USA, Table 5.6 shows a preference at the national (52 %) and site-specific (42 %) scales for transferring it to locations that already manage DOE's defense waste. The second preferred option is different for the national and site-specific groups. National respondents' second preference was moving it to four locations in the USA that would be developed as used fuel storage areas (20 %), whereas the site-specific respondents chose Yucca Mountain (27 %) as their second preference. Only about 10 % favored moving it to a new repository location has not yet been identified.

The site-specific examples are particularly intriguing because all six regions theoretically could be chosen as a place to store this waste under the public's most preferred option. Accordingly, Table 5.6 compares their preference for relocating and storing the waste at defense waste sites versus Yucca Mountain. WIPP-region

Table 5.5 Preference for transferring used fuel to casks as soon as practically achievable, 2011

Site	Choice	Year 2011, %
Site-specific DOE sample ($n = 1{,}080$)	Maintain current policy of letting each separate nuclear facility decide when to transfer used fuel from storage[a]	34.0
	Require transfer of all used fuel from storage pools to concrete casks as soon a practically achievable	55.6
	No response	10.3
National sample ($n = 850$)	Maintain current policy of letting each separate nuclear facility decide when to transfer used fuel from storage	24.5
	Require transfer of all used fuel from storage pools to concrete casks as soon a practically achievable	65.4
	No response	10.1

[a]Site-specific significantly higher than national sample at $p < 0.05$

respondents, who host the newest major nuclear waste management site in the USA, had a stronger preference for relocating the waste at one of these sites (58 %) than any other site or the national sample. The proportion favoring a defense site solution was 51 % at Los Alamos and 43 % at Savannah River. In contrast, less than one-third of Idaho and Hanford site region respondents favored the defense site option. In light of the possibility (see Chap. 2) of expanding WIPP's role in the management of nuclear waste, this is an interesting finding.

Asking if the USA should require transfer all used fuel from storage pools to concrete cast as soon as practically available would require a policy change, and moving the waste from the sites would be a second major policy shift. Such a policy change would require physically moving the waste. Table 5.7 shows public preferences for transportation options. The clear preference is for railroad (57 % national and 53 % DOE sites).

Representing the intersections of the results from Tables 5.1 to 5.7 is a challenge, yet we know that many of the findings just discussed are associated. Factor analysis was used to capture underlying key dimensions. Factor analysis is a multivariate statistical tool that combines highly correlated variables into fewer, new and uncorrelated variables. Table 5.8 shows that factor analysis combined the 19 indicators selected from Tables 5.1 to 5.6 into seven new variables or factors. These seven account for 64 % of the variance of the original 19 variables

The seven factor names in Table 5.8 are derived from the strongest correlations of the original variables with the seven factors. Table 5.8 also provides the proportion of the variance accounted for by each of the factors. The factor loadings in Table 5.8 are the correlations between the original variables and the new factors, which are linear combinations of the original variables. Only correlations of 0.4 or more are shown because highly correlated variables strongly identify with the new factors and lower correlates do not identify with them.

The first factor is called "supporters and opponents of nuclear technology." It contrasts those who are strong supporters of nuclear power even after the Fukushima events with those who already opposed it even before the events. Those who favor increased reliance on nuclear power also favor building new

Table 5.6 Preferred locations for transferring and storing used nuclear waste, 2011

Site	Choice	Year 2011, %
Site-specific DOE sample ($n = 1,080$)	Transfer all used fuel in casks to three or four locations in the USA that already manage and have the US defense waste	42.0
	Transfer all used fuel casks to four locations in the USA that would be developed as used fuel storage areas	19.6
	Move all used fuel concrete cast to a single deep underground repository at Yucca Mountain in Nevada	27.1
	Move all used fuel concrete cast to a single deep underground repository location that has not yet been identified	11.3
National sample ($n = 850$)	Transfer all used fuel in casks to three or four locations in the USA that already manage and have the US defense waste	51.6[a]
	Transfer all used fuel casks to four locations in the USA that would be developed as used fuel storage areas	20.3
	Move all used fuel concrete cast to a single deep underground repository at Yucca Mountain in Nevada	18.1
	Move all used fuel concrete cast to a single deep underground repository location that has not yet been identified	10.0
Hanford ($n = 180$)	Transfer all used fuel in casks to three or four locations in the USA that already manage and have the US defense waste	28.7
	Move all used fuel concrete cast to a single deep underground repository at Yucca Mountain in Nevada	23.0
Idaho ($n = 180$)	Transfer all used fuel in casks to three or four locations in the USA that already manage and have the US defense waste	31.4
	Move all used fuel concrete cast to a single deep underground repository at Yucca Mountain in Nevada	31.4
Los Alamos ($n = 180$)	Transfer all used fuel in casks to three or 4 locations in the USA that already manage and have the US defense waste	50.5
	Move all used fuel concrete cast to a single deep underground repository at Yucca Mountain in Nevada	30.5
Oak Ridge ($n = 180$)	Transfer all used fuel in casks to three or four locations in the USA that already manage and have the US defense waste	38.3
	Move all used fuel concrete cast to a single deep underground repository at Yucca Mountain in Nevada	34.0
Savannah River ($n = 180$)	Transfer all used fuel in casks to three or four locations in the USA that already manage and have the US defense waste	42.7
	Move all used fuel concrete cast to a single deep underground repository at Yucca Mountain in Nevada	20.8
WIPP ($n = 180$)	Transfer all used fuel in casks to three or four locations in the USA that already manage and have the US defense waste	58.1
	Move all used fuel concrete cast to a single deep underground repository at Yucca Mountain in Nevada	22.6

[a]National significantly different from site-specific with chi-square test at $p < 05$

energy-related technologies at DOE's defense sites. In contrast, opponents dispro-portionately are against nuclear power plants and do not want new technologies at the DOE sites, and they also do not want to reconsider nuclear power even though it might help with global climate change as a nonfossil fuel alternative.

Table 5.7 Preferred transportation options, 2011

Site	Choice	Year 2011, %
Site-specific DOE sample ($n = 1,080$)	Truck over interstate highways	25.3
	Railroad	53.4
	Barge over waterways	21.3
National sample ($n = 850$)	Truck over interstate highways	13.9
	Railroad	57.1
	Barge over waterways	29.0

[a]National results are significantly different from site-specific ones at $p < 0.05$

Each of the original 19 variables accounted for 5.3 % of the original variance ($100/19 = 5.26$). Hence, this first factor accounted for 2.6 times as much variance as one of the original variables and is the most important from the statistical perspective, although not necessarily the policy one.

Factor 2 blends together three variables that directly follow from the events in Japan. I call it the "equivocal after Fukushima" factor. It contrasts people who still support nuclear power but are equivocal since the Fukushima events. This set of respondents also is more open to nuclear power because of global climate change, and they do not agree with their counterparts who have not changed their minds about nuclear power because of global climate change. This intriguing set of polar opposite respondents accounted for 11.2 % of the variation (more than double one of the original variables). More analysis is presented below about this equivocal group factor.

The third factor represents pro-fossil fuel and anti-fossil fuel respondents, and the fourth contrasts opponents and proponents of renewable energy technologies. In other words, a distinct group favors increasing reliance on oil, gas, and coal and some oppose all three, and similarly, polarity exists for wind, solar, and biomass energy sources.

The fifth and sixth factors are about managing commercial nuclear waste. Factor five contrasts those who want the used fuel in casks and then moved to a DOE defense site for management. The sixth factor clusters respondents who want the used fuel in casks and then storage in Yucca Mountain. Others did not prefer either of these choices.

The last of the factors contrasted two answers within the same question. This last factor is the weakest statistically (6.5 % compared to 5.3 % for a single variable). It was not analyzed beyond this stage.

5.3.2 Question 2. Correlates of Six Factors

It is unreasonable to expect even the most interested reader to examine scores of regression analyses of the six factors and all the potential correlates. Table 5.9 summarizes the results in a table similar to the one created for Chap. 4, and selected other tabular materials are provided to underscore key observations.

Table 5.8 Seven multivariate components extracted by factor analysis[a]

Component name (% variation in the factor)	Variable	Factor loadings, r-values
Supporters and opponents of nuclear technology (13.7 %)	Have been a firm supporter of increasing nuclear power even after the Fukushima events	0.810
	USA should increase reliance on nuclear energy	0.739
	Firm supporter of nuclear power, irrespective of global climate change	0.729
	Have opposed nuclear power and my concern about global climate change is not changed that opinion	−0.536
	Favor building new energy-related technologies at DOE defense site in my state	0.520
Equivocal after Fukushima (11.2 %)	I am still a supporter of increasing nuclear power but am more concerned than before the events in Japan	0.787
	I have opposed nuclear power and my concern about global climate change has made me more open to considering new nuclear energy facilities	0.777
	Have opposed nuclear power and my concern about global climate change has not changed that position	−0.654
Fossil fuel preferences (9.1 %)	Favor of increasing reliance on oil	0.800
	Favor increasing reliance on coal	0.765
	Favor increasing reliance on natural gas	0.564
Renewable technology (9.1 %)	Favor increasing reliance on wind	0.762
	Favor increasing reliance on solar	0.755
	Favor increasing reliance on biomass	0.578
Preference for moving waste to nuclear defense site (7.9 %)	Favor transporting commercial waste to DOE defense sites	0.865
	Favor requiring the transfer of all used fuel to casks as soon as practically achievable	0.801
Preference for moving waste to Yucca Mountain (6.5 %)	Favor removing all used fuel in casks to a single deep underground repository at Yucca Mountain, Nevada	0.944
	Favor requiring the transfer of all used fuel concrete casks as soon as practically achievable	0.432
Opponents of nuclear power impacted by Fukushima (6.5 %)	Was open-minded about nuclear power, but now against it	0.838
	Have been opposed to nuclear power and global climate change has not changed that opinion	−0.641

[a]Seven components incorporate 64 % of variation from original 19 variables. Results based on a varimax rotation

Table 5.9 Summary results: association between six factors and five sets of correlates

Variable	Support nuclear technology	Equivocal after Fukushima	Prefer fossil fuel	Prefer renewable energy	Prefer moving nuclear waste to nuclear defense site	Prefer moving waste to Yucca Mountain
CONCERN: Less concerned about nuclear power plant in state	Yes[a] (1)	Yes[a](3)			Yes	
Less concerned about nuclear waste site in state	Yes	Yes				
Not concerned about environmental problems	Yes	No[a] (4)	Yes	No		
Environment will be better in 25 years	Yes		Yes	No[a]	No	
TRUST: Trust of responsible parties (scale)[b]	Yes[a] (2)	Yes[a] (1)	Yes		Yes[a] (1)	Yes[a] (2)
DEMOGRAPHICS: Respondent graduated a four year college	Yes		No	Yes		
Respondent is Latino	No[a]	No[a] (2)	Yes[a] (4)	No		
Respondent is Caucasian	Yes		No	Yes[a]		
Respondent is Afro-American	No		Yes[a]			
Respondent is Asian American	No		No[a]			
Respondent is Native American			Yes[a]	No[a]		
Respondent is 55+ years old	Yes		Yes[a] (5)	No[a]		
Respondent annual income exceeds $75,000	Yes		No[a]	Yes[a] (2)	Yes[a]	
Respondent is male	Yes[a] (5)	No[a] (5)			No	Yes[a] (1)
CULTURE & WORLDVIEW: Agree that our society would be better off if the distribution of wealth was more equal	No[a] (4)		No	Yes	Yes	

(continued)

Table 5.9 (continued)

Variable	Support nuclear technology	Equivocal after Fukushima	Prefer fossil fuel	Prefer renewable energy	Prefer moving nuclear waste to nuclear defense site	Prefer moving waste to Yucca Mountain
Agree that discrimination against minorities still a very serious problem in our country	No		No[a]	Yes	Yes[a]	
Agree the government interferes far too much in our everyday lives			Yes[a] (2)		No	
Agree that people who are successful in business have a right to enjoy their wealth as they see fit	Yes	Yes	Yes			
Agree that too many people expect society to do things for them that they should be doing for themselves	Yes[a]	Yes			No	
PERSONAL HISTORIES & PREFERENCES: Agreed that price of gasoline, gas, and electricity influence energy choices		No[a]		Yes	Yes[a] (5)	
Agreed that policymakers should rely on safety and accident history as an influence on the US energy policy	No		No[a]	Yes		
Agreed that policymakers should rely on energy independence as an influence on the US energy policy	Yes[a]			Yes		
Agreed that policymakers should rely on climate change as an influence on the US energy policy	No		No[a] (1)	Yes[a] (1)	Yes[a] (2)	

(continued)

Table 5.9 (continued)

Variable	Support nuclear technology	Equivocal after Fukushima	Prefer fossil fuel	Prefer renewable energy	Prefer moving nuclear waste to nuclear defense site	Prefer moving waste to Yucca Mountain
Agreed that policymakers should rely on waste issues as an influence on the US energy policy			No	Yes		Yes
Agreed that policymakers should rely on environmental considerations as an influence on the US energy policy	No[a]	Yes	No	Yes	Yes	
Agreed that policymakers should rely on human health effects as considerations as an influence on the US energy policy			No	Yes[a] (3)	Yes	Yes
Agreed that policymakers should rely on national security considerations as an influence on the US energy policy		Yes	Yes[a] (3)			
Agreed that policymakers should rely on economic considerations as an influence on the US energy policy	No	Yes	Yes	Yes[a] (4)		
Remember exactly where they were during Fukushima events	Yes[a]	No		Yes		
Experienced a major hazard event	Yes				Yes	
Very familiar with DOE site in state	Yes[a] (3)	No			No	

(continued)

Table 5.9 (continued)

Variable	Support nuclear technology	Equivocal after Fukushima	Prefer fossil fuel	Prefer renewable energy	Prefer moving nuclear waste to nuclear defense site	Prefer moving waste to Yucca Mountain
Very familiar with nuclear power plant in state	Yes[a]				No	
Respondent is Republican	Yes		Yes	No[a] (5)	No	
Respondent is Democrat	No		No	Yes[a]	Yes	
Idaho	Yes	Yes				
Oak Ridge	Yes		Yes[a]	Yes[a]		
WIPP		Yes	Yes			
Savannah River		Yes	Yes[a]	Yes[a]		
Los Alamos						
Hanford						
Multiple r	0.705	0.336	0.554	0.519	0.396	0.332
Number of statistically significant bivariate relationships	30	17	28	23	17	4

Yes means that variables were significant correlates in t-tests or ANOVAs at $p < 0.05$
[a]Correlation was significant at $p < 0.05$ in model in which all variables were entered
[b]Trust scale based on six trust indicators, Cronbach's alpha was 0.894
Number in parenthesis indicates order incorporated into stepwise model, only first five are shown, if 5 or more were found

The dependent variables for the regressions were factor scores derived mathematically from each of the first six factors described in Table 5.8. Factor scores for each factor have a mean value of 0.0. A factor score of 1.0 represents a score 1 standard deviation to the positive side of the average, and one of -1.0 means one standard deviation to the left of the mean. Accordingly, it is relatively easy to interpret factor scores for every respondent.

Each of our respondents has six calculated factor scores. For example, the first respondent of our sample of 1,930 is a 27-year-old white male who has a graduate degree, earned \$75,000–\$99,999, and self-identified as Republican. His factor scores for factors 1–6 were as follows: 0.75, -0.87, -0.60, -0.056, 0.10, and 2.33, respectively. The interpretation of these factors scores is that this respondent is a moderately strong supporter of nuclear technology (FS1 = 0.75). He was not equivocal about nuclear technology (FS2 = -0.87); he is neither a strong supporter of increasing of fossil fuels (FS = -0.60) or renewable energy (-0.56), and he is neutral with regard to moving nuclear waste to DOE defense sites (FS5 = 0.10). He is a very strong supporter of moving used fuel to Yucca Mountain (FS6 = 2.33). In other words, factor analysis allows us to profile each respondent.

Table 5.9 summarizes the major associations between preferences and the five set of correlates. Yes and no in the table indicate a relationship that was statistically significant at $p < 0.05$ based on a one-way analysis of variance or t-test of means or proportions. We then used ordinary least squares linear regression to isolate the strongest predictors. An a in a cell indicates a statistically significant bivariate correlation, and a number in parenthesis shows the order that the specific indicator was incorporated into a stepwise regression. The five strongest correlates are indicated, and in each case, (1) means the first and strongest correlate, (2) the second, and so on. The last two rows are the multiple regression values and the number of bivariate tests out of 40 that were statistically significant at $p \leq 0.05$.

With six factors and 40 predictors, the table has 240 potentially significant associations. By chance, 12 would be statistically significant at $p < 0.05$ ($0.05 \times 240 = 12$). A total of 119, or almost half, actually were statistically significant at $p < 0.05$.

5.3.2.1 Pro and Con Nuclear Energy and New Facilities

The largest number of significant associations (30 of 120 in the first column) and the highest multiple correlation coefficient $r = 0.705$ was with support and opposition to nuclear power and new technology. This was not a surprise. Supporters were less concerned about nuclear-related risks and environmental issues, disproportionately trust responsible parties, and typically are Caucasian, male, highly educated, and relatively affluent. They also are relatively older than their counterparts and identify with the Republican Party, and these respondents assume that the environment will be better in 25 years than it is now. They claim that they are familiar with local nuclear sites. With respect to world views, the pro-nuclear respondents favor less government involvement. The only energy policy that they disproportionately support is energy independence. Respondents from the Idaho and Oak Ridge sites, both with major DOE laboratories, favor nuclear energy and expansion at their locations.

Every factor has a negative and a positives side. Opponents of nuclear technology have polar opposite perceptions, most notably, considerable concern about nuclear technologies, lack of trust of responsible parties, and lack of familiarity with local nuclear sites. Disproportionately, they are Latino, Afro-American, relatively young, poor, less formally educated, and female. They continue to feel that discrimination and concentration of wealth in a small population are issues.

The five strongest correlates of favoring nuclear energy were concern about nuclear energy, trust of responsible parties, familiarity with in-state DOE site, lack of concern about the distribution of wealth in the country, and male gender. Table 5.10 underscores the differences showing average standardized score differences of 1.4 between the opposites for concern, and differences of 0.7 or more for trust, familiarity, and feeling about the distribution of wealth.

5.3.2.2 Pro and Con Fossil Fuels

Factor three had the second highest multiple correlation value ($r = 0.554$) and 28 significant indicators out of a possible 40. This factor highlights the differences

Table 5.10 Correlates of support and opposition to nuclear technology

Variable	Commercial nuclear power plant average factor score (n = number of respondents)
Worry about local nuclear power plant (1–10)	
1 (not at all worried)	0.94 (331)
2	0.57 (153)
3	0.31 (129)
⋮	⋮
8	−0.41 (72)
9	−0.68 (40)
10 (extremely worried)	−0.46 (88)
Trust of managers	
1 (most trusting)	0.25 (591)
2	0.18 (219)
3	0.07 (406)
4 (least trusting)	−0.44 (468)
Familiarity with local commercial nuclear power site	
Very familiar	0.65 (291)
Somewhat familiar	0.12 (495)
Only a little familiar	−0.21 (1083)
Agree that our society would be better off if the distribution of wealth was more equal	
Strongly agree	−0.25 (661)
Somewhat agree	−0.21 (455)
Neither agree or disagree	−0.02 (45)
Somewhat disagree	0.09 (293)
Agree	0.58 (436)
Respondent is male	
Yes	0.24 (944)
No	−0.23 (986)

[a]Average factor score values are 0.0

between supporters and opponents of fossil fuel use for generating electricity. Proponents of fossil fuel use do not think that the US government should rely on climate change as an influence on energy policy, but that they should rely on national security considerations, and these pro-fossil fuel respondents perceive that the government interferes too much in our everyday lives. Disproportionately they are Latino, Afro-American and Native American, and 55+ years old. They are relatively less formally educated and less affluent and are concerned about the personal cost of energy, and economic considerations are more important to them than environmental and public health ones.

Their counterparts have the opposite attributes, especially the view that climate change is a critical policy issue, that the government does not interfere too much in our everyday lives, and these respondents are disproportionately younger and Caucasian.

The most striking difference was observed for the climate change as a policy factor variable. Those who strongly disagreed with the idea of increasing fossil fuel use ($n = 909$) had an average factor score of -0.14 compared to 0.57 among the 157 respondents who strongly concurred with the idea of using fossil fuel use.

5.3.2.3   Pro and Con Renewables

Preferences for renewable energy sources was the third strongest factor, measured by a multiple r-value of 0.519 and 22 out of 40 possible statistically significant correlations. These respondents wanted energy policy to be strongly influenced by global climate change, by human health, as well as economic considerations. They self-identified as relatively well educated, Caucasian, and Democrat. However, reaction to using climate change as a policy consideration was the key predictor. Those relatively few ($n = 157$) felt that climate change should not at all be a big consideration in federal policy manifested an average standardized score of -0.75 with the renewable energy factor, and those who responded "not much" ($n = 124$) had an average score of -0.53. In contrast, the almost 1,600 respondents who wanted climate change to be a key policy consideration had an average score of 0.13 with the renewable energy factor.

Factors 1, 3, and 4 were statistically stronger than were factors 2, 5, and 6. Factors 2, 5, and 6 are less well defined by the statistical analyses, yet they follow from the Fukushima events and consequently merit special attention.

5.3.2.4   Fukushima Impacted

Factor 2 is strongly Fukushima influenced, contrasting those who are supporters of nuclear power but have become equivocal as a result of the event and are more open to nuclear power because of global climate change with those who were and continue to be opposed. Seventeen statistically significant correlates out of a possible 40 were found. But these send a clear message. The equivocal but supportive group trusts responsible parties and is not very concerned about the nuclear plant(s) or DOE nuclear facility in their state. The striking finding is that these respondents are disproportionately women, not familiar with their local DOE site, and are concerned about environmental issues.

The key variable driving this relationship is trust. That single trust variable had an r-value of 0.233. Because trust was a key predictor in this and factors 5 and 6, the six original trust indicators were correlated with the factor scores. With regard to factor 2, equivocal post-Fukushima respondent, the r-value rose from 0.233 to 0.307. More important, two trust indicators were identified as significant: (1) trust the DOE to effectively manage any new nuclear-related activities and (2) the DOE will make sure that underground radioactive and chemical materials at the site do not pollute the air, land, and water outside of the site boundaries. The almost 1,200 respondents who trusted the DOE to safely manage new on-site activities had an average factor score with factor 2 of 0.15 compared to -0.36 for the over 600 individuals who did not trust DOE. Similarly, with regard to preventing off-site leakage of existing wastes, those who trusted the DOE to effectively manage had an average factor score of 0.16 compared to −0.34 for those who did not trust the DOE to control existing on-site wastes. In short, it appears from these data that trust of the DOE is a major reason why a large proportion of these equivocal respondents continue to feel positively about nuclear energy. (We did not ask about the NRC in the 2011 survey because earlier surveys showed that the vast majority of the public did not understand that distinction. In hindsight we should have.)

5.3.2.5   Moving Commercial Waste to Defense Sites and to Yucca

The six-variable trust scale was the strongest or second strongest correlate of these factors, and yet, the relationship using the full trust scale was misleading. Once the results were split between DOE and private contractors, we found that respondents who favored moving commercial nuclear waste to defense sites and to Yucca had lower trust of contractors, and they were neutral with regard to trust of the DOE (see Table 5.12 below for further presentation of DOE results). With regard to contractors, proponents of moving casks to defense sites idea disproportionately did not trust contractors to safety mange new nuclear activities, and to make sure that underground radioactive materials will not escape the site. Perception of contactors surely is an interesting issue with regard to defense site option.

Yet these results should not be taken at face value because the trust questions were not designed to dig deeply into this potentially policy-relevant observation. Arguably, the respondents are saying that distrust of private operators of nuclear power plants makes them more trusting of the DOE defense site waste option and of Yucca, or the lesser of two evil preference. In order to better understand the meaning of this finding, a much more finely scaled set of questions is needed that would discuss this face-to-face in a focus group setting or a survey that had many more questions about the issue. Given the small number of questions, I would call the evidence to support this government-private dichotomy in trust as a driver of public preference underwhelming and yet certainly worth follow-up.

5.3.2.6   Trust, Global Climate Change, and Fukushima

The consistent important role of trust in the post-Fukushima data analysis is the single most noteworthy new finding from the CRESP 2011 survey. Table 5.11 summarizes public responses to the same six trust questions in 2010 and 2011. In 2010, 60–76 % of respondents trusted the DOE and contractors to manage existing waste materials, communicate honestly with the public, and effectively manage new nuclear-related activities. In 2011, the numbers dropped to 53–71 %. The largest declines were in managing existing wastes among both DOE and contractors. The largest change was from strongly agreed that they trust the responsible parties to disagree. Both also had reduced proportions for communicating honestly. The smallest drop was for effectively managing new site activities, which slipped 4–7 % from strongly agree to agree.

Table 5.11 Trust for site parties at DOE sites, 2010 and 2011

Question	Choice	Site-specific sample, 2010	Site-specific sample, 2011
DOE will makes sure that underground radioactive and chemical materials at the site do not pollute the air, land and water outside of the site's boundaries ($n = 2,014$ in 2010, 1,327 in 2011)	Strongly agree	26.2[a]	14.9
	Agree	50.2	50.8
	Disagree	19.4	29.4
	Neither or no response	4.2	4.8
Site contractors will make sure that underground radioactive and chemical materials at the site do not pollute the air, land, and water outside of the site's boundaries ($n = 2,003$ in 2010, 1,314 in 2011)	Strongly agree	24.9	16.3
	Agree	47.5	49.2
	Disagree	23.0	30.4
	Neither or no response	4.5	4.1
DOE communicates honestly with the people in the site's area ($n = 1,915$ in 2010, 1,329 in 2011)	Strongly agree	19.1[a]	10.4
	Agree	44.6	43.0
	Disagree	27.5	41.5
	Neither or no response	8.8	5.1
Site contractors communicate honestly with the people in the site's area ($n = 1,908$ in 2010, 1,310 in 2011)	Strongly agree	18.2	10.9
	Agree	42.0	43.6
	Disagree	30.6	38.6
	Neither or no response	9.1	6.9
I trust the DOE to effectively manage any new nuclear-related activities ($n = 2,010$ in 2010, 1,370 in 2011)	Strongly agree	17.5[a]	10.7
	Agree	54.2	60.0
	Disagree	24.1	26.3
	Neither or no response	4.3	3.0
I trust contractors to effectively manage any new nuclear-related activities ($n = 2,005$ in 2010, 1,340 in 2011)	Strongly agree	13.9	9.5
	Agree	50.0	53.1
	Disagree	31.7	32.4
	Neither or no response	4.5	4.9

[a]Paired samples t-tests show that in 2010 DOE is significantly more trusted than contractors at $p < 0.05$

This chapter ends by highlighting two especially intriguing findings. Public support for nuclear energy and new nuclear activities at DOE sites post-Fukushima appears to depend heavily on perceptions of trust and global climate change. A final demonstration follows. As noted in Table 5.1, 43 % of all respondents continued to support nuclear energy but were more concerned after the events in Japan. The strongest intersection of this preference is with concerns about global climate change. Thirty-eight percent of respondents said that global climate change had made them open to considering new nuclear energy facilities (Table 5.2). In fact, 60 % of those who said that they still supported nuclear energy but had become more concerned after Fukushima came from this global climate change group.

The second piece of the equivocal nuclear support, global climate change concern, and DOE trust triangle is illustrated by the observation that those who were supportive but more concerned about nuclear power said that they "agreed" that DOE could effectively manage any new nuclear-related activities. Specifically, 65 % of those who said that remained supporters but were more concerned after Fukushima also indicated that they agreed that DOE could manage new nuclear facilities. However, those in this group were underrepresented in the strongly agree, disagree, and strongly disagree groups. In other words, they were not willing to say that had near complete confidence in the DOE to manage new facilities (that choice was strongly agree with trust), nor were they willing to say that they did not have confidence in the DOE, their trust was lukewarm. Joining the people who said that they "agreed" that DOE could effectively manage new nuclear sites and had become open-minded about nuclear energy because of concern about global climate change produced a group of over 400 respondents (out of 1930). Seventy-one percent of this group was in the still supported nuclear energy but was more concerned after Fukushima group. This compared with 34 % of all other survey respondents. In essence, moderate trust of a key federal agency and concern about global climate change appear to be two reasons why post-Fukushima support for nuclear power did not dramatically decrease.

With regard to DOE's missions, however, it would be misleading to overstate the importance of only trust. Table 5.12 shows a strong relationship between trust of DOE and willingness to support new on-site missions. Yet, there is no relationship between trust of DOE and two interesting decisions that would involve the DOE: movement of used fuel to DOE sites for "temporary" storage or movement of it to Yucca Mountain. The last two policy changes may be too disconnected from the current realities, which implies that there should be time to develop and interact about the options, while building trust in the responsible parties. The immediate driver there appears to be distrust of contractors rather trust of DOE.

5.4   Summary and Lessons Learned

We summarize the changes made for this survey, the limitations, and results of the 2011 survey following the format of Chap. 4.

5.4.1   Design Changes and Limitations

L1. We added a cell phone component of 25 % in order to avoid potential bias associated with landline only surveys. The cell phone sample was for each individual site as well as a national survey. Response rates were similar to those in our recent surveys.

L2. Sample sizes for the six sites were 180 (a total of 1,080 respondents), and the national sample size was 850.

Table 5.12 Trust of DOE and selected public preferences

2011 Questions	Support or Oppose the Decision	New on-site activities[a]			Relocate commercial used fuel to DOE defense sites			Move Used Nuclear materials to Yucca Mountain		
		SA	AG	Don't	SA	AG	Don't	SA	AG	Don't
DOE will makes sure that underground radioactive and chemical materials at the site do not pollute the air, land, and water outside of the site's boundaries	Support	0.22	0.50	0.28	0.14	0.47	0.39	0.13	0.48	0.39
	Do not support	0.12	0.47	0.41	0.15	0.48	0.37	0.15	0.48	0.37
DOE communicates honestly with the people in the site's area	Support	0.15	0.42	0.43	0.10	0.43	0.47	0.08	0.37	0.55
	Do not support	0.08	0.41	0.49	0.10	0.41	0.49	0.11	0.42	0.47
I trust the DOE to effectively manage any new nuclear-related activities	Support	0.17	0.59	0.24	0.09	0.57	0.34	0.11	0.47	0.42
	Do not support	0.07	0.54	0.39	0.11	0.55	0.34	0.10	0.57	0.33

[a]Paired samples t-tests show that in 2010 DOE is significantly more trusted than contractors at $p < 0.05$

L3. Survey length was 23 min, which limited the numbers of questions but means that responses should have minimal respondent fatigue.

5.4.2   Results: New Activities at Legacy Sites

R1. Preferences for new nuclear-related activities in the respondents' host state dropped from 48 % in 2010 to 33 % in 2011. Everyone of the sites showed a substantial decrease.

R2. The Idaho site, host county residents, and those who are familiar with the site because they worked there or had a relative who did continued to be the strongest supporters.

5.4.3   Results: Impact of Fukushima Events on Preferences for Nuclear Power

R3. Almost 60 % of national respondents and 73 % site-specific ones favored nuclear power. However, about two-thirds of these were more concerned about nuclear power than before the events in Japan. Between 2010 and 2011, the proportions favoring greater reliance on nuclear power for electrical energy decreased approximately 15 %, whereas the proportion favoring natural gas increased almost 10 %.

R4. In 2010, 38 % of respondents said that global climate change made them more open to considering nuclear power and this proportion did not change after the Fukushima events.

5.4.4 Results: Legacy Waste Management

R5. Approximately 60 % of respondents prefer the transfer of use fuel from storage pools to concrete casks as soon as practically achievable. About half prefer transferring the used fuel in casks to already existing US defense waste sites and about 40 % prefer transferring it to either Yucca Mountain or four locations in the USA that would be developed as used fuel storage areas. Over half of WIPP and Los Alamos respondents favored moving the waste to defense waste sites compared to about 30 % of Hanford and Idaho site region respondents.

R6. Over half of respondents preferred moving this waste by railroad compared to about 25 % by barge and waterways and less than 20 % by truck over interstate highways.

5.4.5 Results: Correlates

R7. Concerns about safety and exposures, and associated emotions and feelings were the strongest correlates of support for nuclear technology. Indicators of worrying and concern were also important in helping to explain public reactions to the Fukushima events.

R8. In 2010, trust was relatively high, averaging about 70 % for federal agencies and contractors. One year later, after Fukushima, trust fell approximately 10 % and the strongest trust category but remained over 60 %. Trust of responsible parties was the strongest or second strongest correlate of many of the results. Most notably, trust of DOE is a very strong correlate in the post-Fukushima data.

R9. The so-called white male effect with affluent college-educated white males disproportionately supporting new on-site activities was observed in the 2011 data. In a number of these analyses, the advantages of affluence, education, and male gender in gaining access to power or at least the feeling of it were reinforced by cultural and worldviews that supported individualism and hierarchical power structures.

R10. Other factors, most notably personal familiarity, feelings about the relative importance of environmental protection, human health, global climate change, national security, and individual and societal economic benefits, were important correlates of preferences for nuclear power, other forms of energy, and post-Fukushima ambivalence about nuclear power.

5.4.6 Policy-Related Results

R11. Prior to Fukushima, there was considerable local public support for new nuclear missions and intensifications of existing ones. After the events in Japan, support for in-state expansion of nuclear activities at DOE defense sites dropped by 10–18 % at every site. However, support for expansion of nuclear-related activities at Idaho, Oak Ridge, and Hanford remained high at one-third or higher.

R12. Increased preference for the use of natural gas substantially increased between 2008 and 2011, which will be a challenge to not only the DOE but agencies such as FERC, EPA, and others that are charged with managing LNG facilities and recovery of natural gas from environment.

R13. Fukushima introduced the US residents to the used fuel issue and reopened the issue of long-term management of commercial nuclear waste. Public preference would require substantial policy adjustment in regard to the management of used fuel, where it would be managed over the long-term, and how it would be transported for long-term management.

R14. A new strong dynamic in the data is the appearance of a large group that supports nuclear energy and new nuclear-related facilities but has become more equivocal in their support since Fukushima. This group of respondents disproportionately has become more open to nuclear energy because of global climate change, and they trust DOE but that trust is not at the highest level. This group, specifically, is an important one to address if the DOE, NRC, and other pro-nuclear industry groups want to maintain their support.

Indeed, the data show that both the federal agencies and for-profit managers of power plants and waste management sites have suffered a moderate loss of credibility because of the events in Japan. The multiple publics with varying backgrounds and issues need to be sought out by the responsible parties in order to address Fukushima-related concerns and to maintain credibility that will be required to safely manage the defense and commercial nuclear industries.

References

Blumberg S, Luke J (2010) Wireless substation: Early release of estimates from the National Health Interview Survey < January-June 2010. National Center for Health Statistics, USA. http://www.CDC.gov/nchs/NHIS.htm.2010. Accessed August 24, 2011

CTIA, the Wireless Association (2011) Wireless quick facts. http://www.ctia.org/advocacy/research/index.cfm/aid/10323. Accessed August 5, 2011

Ipsos (2011) Global citizen reaction to the Fukushima nuclear plant disaster. PowerPoint presentation. http://www.ipsos-mori.com/rsearchpublciations/rsearcharchive/2817. Accessed September 29, 2011

Pew Research Center for the People and the Press (2011a) Opposition to nuclear power rises amid Japanese crisis. Http//People-press.org/2011/03/21. Accessed September 19, 2011

Pew Research Center for the People & the Press (2011b) Japanese resilient, but see economic challenges ahead. http://www.pewglobal/org/2011/06/01/Japanese-resilient-. Accessed September 19, 2011

Chapter 6
Nuclear Waste Management: Building a Foundation to Enhance Trust

Abstract The primary challenges to the DOE, NRC, EPA, and contractors responsible for managing the defense and commercial nuclear waste legacies are technology and high cost. Yet given the long period that the waste must be stewarded, the responsible parties must invest in short-term and long-run programs to build mutually beneficial stable relationships that are able to withstand the stresses of technical mishaps and human disagreements and endure for many generations. This chapter suggests five key steps in light of the Fukushima events:

1. Do not shoot the messenger. Fear needs to be acknowledged, not dismissed as irrational.
2. If the audience wants to understand what happened in Japan, or in other incidents involving nuclear facilities, then engage in a discussion if you feel competent to do so, or try to find someone who can.
3. People want the responsible parties to demonstrate ability to protect them, and prove through these actions that safety is the highest priority now and in the future.
4. The public wants promises about definitive steps to improve safety, efficiency, and other metrics of competence; they want to know what communications will keep them or their representatives in the loop; and they want to be sure that the safety is not going to be sacrificed.
5. Follow-through with promised actions. Trying to change public opinion with words not matched by deeds will erode, perhaps even poison, what could be a productive ongoing organizational relationship.

More detail is provided in the body of the chapter.

M.R. Greenberg, *Nuclear Waste Management, Nuclear Power and Energy Choices*,
Lecture Notes in Energy 2, DOI 10.1007/978-1-4471-4231-7_6,
© Springer-Verlag London 2013

6.1 Introduction

After conducting five surveys, writing more than a dozen papers and this book about the subject of public preferences and perceptions associated with nuclear waste management and nuclear energy, I hope that our surveys have contributed to public discourse by providing clues to government and private decision-makers about future directions. The purpose of this chapter is to summarize these contributions and to offer some modest suggestions.

6.2 Public Presentations: An Opportunity

I have spoken more than 50 times in public about the subject matter of this book and related ones. My objectives have been to provide information and respond to questions about public preferences for waste management practices, nuclear and other forms of energy, and US energy policies. I have tried to provide that information without introducing my personal feelings and preferences or at least stating them at the onset. I continue to receive speaking invitations, which I take as a good sign.

The risk communication literature recognizes that communication is a two-way process (Committee on Risk Perception and Communication 1989; Stern and Fineberg 1996), and I have learned from these sessions. One unanticipated observation was about public opinion polling. Some attendees are slightly familiar with the science of public opinion polling, for example, they know that random sampling is involved, even though they are not exactly sure what that is. I learned that some are interested in learning more about public opinion polling, beyond its application to energy and waste management issues. Here is an illustration. A PhD scientist told me and about 100 other people that he could save the taxpayers a great deal of money (in other words, I was wasting taxpayer money by polling) by going to his favorite big box retailer, passing out a survey, and paying each respondent $10. My response was that his big box store would provide a sample of older upper middle-income Caucasian males from one part of the country (no DOE sites within hundreds of miles). His convenience sample would not be close to a representative sample and was not random. His question and my responses were eye-openers for members of that audience.

Because reporters often attend these public meetings, a second benefit is helping them understand that not all surveys are equally credible, which I have found to be a problem when reporters try to "balance" coverage by giving equal space to both sides. In short, the public is bombarded by survey information. Those who understand the science behind surveys should take opportunities to explain that there are differences among surveys and provide some clues about how the public can judge the credibility of results for themselves. I never anticipated making such a presentation, and now I come to meetings expecting to for a small portion of the audience.

A second lesson learned is not everyone is math and science-phobic people. Most have no tolerance for long tables and equations. But my experience is that many want to get past slogans by interest groups on both sides of these issues. Indeed, for example, they expect some organizations to portray the impact of the Fukushima event in March 2011 as relatively minor and others to argue that the public no longer supports nuclear energy. Even if they use the CRESP data to support their preexisting position (which I have noticed), I have found that there are always people who want more details and are comfortable looking at up to three tables. For example, I have a reduced version of the table in Chap. 5 that shows the change in trust after Fukushima, a second table that shows the number of people who still support nuclear power after Fukushima but are now more concerned, and a final table that shows how this ambivalent group primarily is composed of those who still trust the responsible parties and also are worried about global change. This last table validates the feelings that many in the audience have, which I see when I look up and see nods and respond to questions. I am not talking about presenting linear or logit regression models to audiences, but I am talking about an opportunity for the numbers to speak for themselves in ways that build credibility rather than serve to separate the scientists from the public.

6.3   Decision-Makers and the Public: Building a Foundation

Chapter 3 showed that a great deal already has been written about public preferences and perceptions about nuclear and other forms of energy. Much less has been written about nuclear waste management, and very little of it is from the vantage point of those who live near nuclear waste management facilities. The major contribution of the CRESP work has been to make nuclear waste management and the major DOE waste sites a focus, not an afterthought that needs to be discussed elsewhere. To reiterate a point first made in Chap. 1, it does not matter if nuclear power plants make only a marginal contribution to growing US electrical energy demand or if two dozen plants are constructed. The defense nuclear waste legacy will still be there, and the current nuclear waste will need to be managed. If there is a substantial increase in reliance on nuclear energy, there will be more waste to manage. But the current legacy is quite a challenge, and the responsibility for managing it will not disappear whether there is a nuclear renaissance or not. This section summarizes insights derived from the CRESP studies in three sections: nuclear waste management, nuclear power, and building trust for long-term stewardship.

6.3.1   Nuclear Waste Management Issues

Fukushima raised the stakes for both the DOE and the NRC. But they were already high when in January 2010, the president of the USA appointed a Blue Ribbon Commission to look at the nuclear waste management issue, especially commercial nuclear waste. On January 31, 2012, the commission offered multiple recommendations (Blue Ribbon Commission 2012). Paraphrasing, they said that Congress should:

1. Create a corporation with a charter that will site, license, build, and operate nuclear waste management and storage facilities
2. Develop a facility-siting process grounded in cooperation with states, tribal nations, and local governments
3. Amend federal law so that the $27 billion nuclear waste fund can be used for nuclear waste management
4. Begin to look for a permanent repository, regardless of the ultimate fate of the partially completed Yucca Mountain one
5. Amend the law to allow for the identification of one or more "consolidated" storage facilities
6. Prepare for shipping used nuclear fuel and high level waste to consolidated storage and disposal sites
7. Support research and development of technology and education of a nuclear labor force
8. Continue to be heavily involved in assisting countries with development of nuclear programs

These recommendations are not knew ideas; I heard almost all of them before 1980. The committee organized them in a single coherent presentation and most important urged action. It is not my intent to agree or disagree with the panel's recommendations in this book. It would be a diversion of the objectives of the book. The commission's recommendations are consistent with a view, which I share, that we are not facing a public health or economic emergency. The DOE has recognized a moral and legal commitment to manage the high level waste from the bomb making processes and the commercial waste industry, albeit the path for the latter is more difficult to predict (see Chap. 2). The DOE has been implementing an accelerated site closure strategy, trying to close sites that no longer have missions, implementing project management approaches to reduce uncertainty and costs, and building long-term stewardship plans (Office of Environmental Management 2000 and Chap. 2). Arguably, given the tight budget of the US government, we should expect budget reductions for the DOE major waste management programs at the Hanford site and to a lesser extent several others. We know that the American Economic Recovery Act of 2009 provided DOE with $35.2, including $6 billion for environmental management (US Department of Energy 2012), which allowed the DOE to complete projects and accelerate progress on others.

Funds for remediation are going to decline. David Huizenga (2012), senior advisor for environmental management for DOE, summarized the DOE's accomplishments by noting that it began operations in 1989 with responsibility for 110 sites in 35 states and well over 3,000 square miles of land. At the end of 2011, the legacy was 17 sites in 11 states. Huizenga's report estimates that the DOE will need another $200 billion to complete the cleanup mission. The DOE report shows funds dropping in the immediate future. The implication is that the EM program will have to make tough choices among the tank wastes, spent fuel, and other components of its program. Notably, 83 percent of the 2011 and 2012 budgets were allocated to the sites in our surveys (Hanford, Idaho, Los Alamos, Oak Ridge, Savannah River, and WIPP). Barring an unforeseen economic collapse, it is hard to conceive of the federal government neglecting the nuclear and chemical waste legacy; rather, it would postpone meeting some milestones.

DOE, NRC, and EPA's involvement in overseeing management of the waste from defense and commercial nuclear power plants remains, but some important changes could occur. This author does not see any alternative to "temporary" storage at the existing operating nuclear power plants. The BRC noted that it might take a decade to find what it called consolidated storage sites. Meanwhile, some existing science and technology policies are being reconsidered. One is not locating facilities in areas with cataclysmic natural hazard potential. The Japanese plants would also have been much less damaged if viable and sustainable emergency power sources had been available. This failure strongly suggests the need for revisiting backup power supplies at nuclear plants. A third challenge is what to do with so-called "used" fuel. The events in Japan alerted many US residents that spent fuel has almost all of its energy intact. Leaving the used fuel in large swimming pools is arguable. We know that this means transport of nuclear materials, which our surveys show is a major public concern, but the Japanese nuclear reality suggests the need to reassess our policy of leaving spent fuel at over 60 US sites for decades. Notably, several states, including NJ, have just challenged the NRC decision to store spent fuel for additional decades beyond the initial 30 years.

Our survey data address these and several other options. First, the CRESP survey data show that those who live near the major DOE sites have high expectations of the federal agencies and contractors about managing the waste sites, especially regarding environmental surveillance. The public wanted an extremely active monitoring program before the events in Japan. I cannot imagine that they now want a reduced one. Second, the public, especially the public that lives near the DOE sites, is not adamantly opposed to "temporarily" managing the commercial nuclear waste in their backyards, especially WIPP. Fifty-eight percent of WIPP respondents favored moving waste in canisters to a defense site, which was 7.5 % higher than Los Alamos, the second highest supporter (see Table 5.6). I do not think that this result is independent of the reality that WIPP handles transuranic wastes and that the siting process for WIPP involved considerable input from the state, local groups, and individuals. Considerable trust was built up during the WIPP siting process, which seems to carry over to the survey results and is consistent with a study by Jenkins-Smith et al. (2011 and Chap. 3).

Third, the public is concerned about transporting the waste and strongly prefers rail transport. Fourth, now that the Fukushima event has made them aware of "used" nuclear fuel, they have expressed a clear preference for storing it as soon as possible in casks rather than leaving it in pools, and then transporting it from commercial nuclear plants to DOE defense sites. Our 2011 survey, which asked these questions, clearly needs to be repeated with multiple probes about these issues. To be perfectly clear, multiple additional surveys are required to replicate the findings of the 2011 survey. There is time to conduct these studies. Another piece of our data is less preference for finding new sites. That is, defense sites were preferred to new temporary locations, and Yucca was desired more than a new permanently waste site.

6.3.2 Nuclear Power in the Energy Mix, Deliberation, and Trust

The images we saw from Japan were painful. I feel great sorrow for Japan as a nation, and especially for those who lost family members, their communities, and personal possessions. Yet Japan is thousands of miles away, and the USA is afforded the opportunity to deliberate about the implications of the Japanese nuclear failures in the US context, especially those discussed under commercial nuclear waste management.

I think that the most important policy issue for new nuclear plants is what will happen to financing of new nuclear plants. The US government has provided loan guarantees to companies willing to construct new facilities and went on record as favoring more loans. When a company has billions of dollars in loan guarantees to support their project, they face a less risky economic climate. How will banks and other private financial institutions respond to the possibility of no loan guarantees? Will the US Congress see or use the Japan events as a reason to remove loan guarantees? How will smaller utilities build plants without federal government financial support or other guarantees?

Our surveys had no questions directly about this important issue. Nor did we have questions about relicensing older nuclear power plants. Quite a few of our existing plants are not very different from those in Japan. In light of events in Japan, the policy to relicense "old nukes" is being scrutinized, and presumably opponents of nuclear power will use public opinion about declining support for nuclear power in their arguments. On the other side, proponents will argue that the plants are safe and that closing nuclear plants will increase energy costs and require more use of gas, coal, and oil. What could elected officials take from these surveys?

The CRESP data show that the US public was divided in support for greater reliance on nuclear power and remains so after the Fukushima events. Hence, neither side of the endless nuclear power debate has won a decisive victory in the USA, irrespective of some of the headlines that do not seem to reflect the data they actually present or partially present. The most important piece of information we added was the overall decrease in support and an increase in support conditioned on concern about global climate change and trust of federal agencies and contractors that supervise the sites. Support for nuclear power was higher in the DOE site areas before Fukushima, and while it declined, remains higher as does trust of DOE.

Those who advocate for the demise of nuclear energy need to consider that many US states from across the country have policies to reduce energy use and dependence on fossil fuels, and in some of these local nuclear power plants are part of the mix. Alternatively, these states would need to import electrical energy from other locations, which suggests a loss of local control, higher costs, and more air, water, and land pollution in-state or elsewhere. For example, obtaining more natural gas from fossil fuel containing shale and building new coal- or oil-fired facilities are all controversial options. The events in Japan could accelerate efforts to increase our use of solar, wind, and other energy sources (e.g., hydro, geothermal, bio). But the first two only accounts for about 2 % of our supply, and it is not easy to quickly increase the supply from these facilities to compensate. There is also the reality that the public is now becoming aware of the implications of natural gas, liquefied natural gas, transmission lines, large clusters of wind and solar facilities, and the towers that link them to the grid system. The only nonnuclear alternatives that supply large amounts of energy needed to run our computers, homes, and appliances in many states are coal and natural gas, and, of course, conservation which would help but requires the public to embrace the option and personally sacrifice.

With regard to electrical energy sources, the CRESP survey results are similar to those of others (Chap. 3). That is, solar and wind power are the public's favorite, followed closely by natural gas, and then nuclear power. Coal and oil are farther down the preference list. The CRESP data added the key observation that the public is aware of fuel cycles. To reiterate (Greenberg and Truelove 2011), pre-Fukushima nuclear power plant accidents were not the strongest predictor of opposition to greater reliance on nuclear power. Poor management of nuclear waste products was the stronger predictor. And global climate change was not the strongest predictor of opposition to coal use for electricity; coal mining was.

6.4   The Public Participation Challenge and Public Trust

Reading and listening to journalists' and advocacy group questioning of nuclear experts post-Fukushima, watching news shows, and reading articles could lead the public to believe that the Nuclear Regulatory Commission, DOE, and others in charge of commercial and defense nuclear facilities in the USA are not capable of managing nuclear materials. If the events in Japan lead the public in our area to question the commitment of responsible parties to safely manage the plants, then the nuclear power industry could be in deep trouble in keeping open existing plants and building new ones. The DOE's ability to manage their legacy wastes has been challenged but primarily on the grounds that they have failed to move forward on a permanent storage site and to effectively spend resources collected for that purpose (see Chaps. 2 and 5, and the eight recommendations of the BRC above).

Frankly, the public appears to be more tolerant of the DOE than Congress and business. The CRESP surveys show that the public disproportionately trusts the DOE and companies to manage the sites and keep the public informed. Both the DOE site regions and the USA as a whole manifested a decrease in support for nuclear power after Fukushima, but as noted above, I think the decline was not as substantial as headlines in some media suggest. The expected predictors (worry, trust, worldviews/culture, personal experience, and demographic attributes) as predicted help us understand the variation in preferences and perceptions. In the pre-Fukushima surveys, worry measures and indicators of affect clearly were the strongest predictors. After the event, the importance of trust and worry about global climate change rose in importance. Overall, the CRESP data suggest that there is no crisis in public confidence in the US nuclear industry.

Before offering some modest recommendations about the importance of building trust, it would be disingenuous to ignore that some have told me that efforts to survey and actively engage the publics are a waste of time because as noted above the government has a moral and legal obligation to manage the sites and would demonstrate inconceivable stupidity to step away from their obligation, even if the public wanted it to. Furthermore, I am reminded that local communities have a financial incentive to remind the federal government of their obligations and their state governments an incentive to press the federal government to continue to invest in the legacy wastes and use the sites for new missions like energy parks where a broader spectrum of energy research could be undertaken. In other words, the local governments and their state representatives will not allow the DOE, NRC, and EPA to withdraw. On the private side, the Nuclear Energy Institute and its clients from all over the world have a much greater stake in maintaining public support in the USA, especially because Germany and Japan have publicly stated that they are phasing out nuclear power. The producers of nuclear energy could lose their investments and jobs, and so they have a near desperate need to build public trust.

6.4.1 Five Publics

Engaging the public is a challenge because there is no single public with regard to the issues discussed in this book. The CRESP data have allowed us to identify five US publics about energy preferences and waste management, and I think some of our five groups contain subgroups (Greenberg et al. 2011; see also Leiserowitz et al. 2009; Miller and Sinclair 2012).

I briefly describe them because the existence of multiple groups has implications for building trust and participation. One group, the "supporters," is disproportionately a high socioeconomic status group of Caucasian males 45 years and older who are focused on self and favor hierarchical organizational management. This group favors greater reliance and nuclear power and is largely comfortable with DOE's waste management practices. It strongly opposes coal and oil as an electrical energy source. The group trusts the DOE, NRC, EPA, their state governments, and the contractors. They are willing to embrace new nuclear facilities in their area and welcome energy parks. Comprising about 5–30 % of DOE site region populations, they are familiar with the sites, are more knowledgeable about nuclear facts, and typically are the ones who have feelings and images of jobs, income, pride, beautiful, mountains, and others associated with the DOE sites.

On the opposite side of the coin are the "opponents," who are relatively well educated, typically relatively young Caucasian women who oppose nuclear energy, oppose new facilities in their region, and are not satisfied with DOE waste management practices, and they also opposed fossil fuel energy sources. They strongly favor solar and wind and, in general, are much more concerned about environmental quality than economic growth now and in the future, and they comprise 10–40 % of DOE site region populations. Notably, my personal experience is that opponents and the supporters disproportionately tend to go to meetings, which as noted below complicates the process of building sustainable trust.

A third group, typically the largest (25–60 %), is mostly young adults who are not aware of many of the issues about nuclear waste and power and they are not very interested in learning about them. What they do know, or think they know, is often inaccurate. I call them the "disinterested" group.

The fourth measureable group is only about 2–10 % of the population and consists of relatively poor, primarily minority, and older respondents who appear to be a pro-fossil fuel population without a strong view of nuclear power, but lacking in trust of government officials. They are much more concerned than others about the cost of all forms of energy. It is hard to label them, but I have chosen "cost-conscious" group.

The fifth group cannot be found in surveys such as those in this book. It consists of economic and political elites who disproportionately can reach out to elected officials and businesses in support of or in opposition to the actions. They are frequently at public meetings. I call them the "stealth" group because they tend to be present but not participatory in public settings.

6.4.2 Trust and the Five Publics

There are multiple definitions of trust. For purposes of this book, when people believe that an organization and its employees are competent, communicate honestly and often enough with them, and share many of their key values that organization will be trusted, at least until a clear problem occurs to change their mental model of the organization and its people. Chapters 3–5 illustrated the significance of trust.

Responsible US agencies, utilities, and contractors took a moderate hit in trust, even though they did not cause the events in Japan. Rather than feel inappropriately under siege, nuclear industry professionals need to accept that reality that their position is similar to every airline when another airline has a crash, every department of transportation when a bridge collapses in a another state, every investment firm after some recent disclosures, and so on for every profession.

My five post-Fukushima suggestions for the nuclear industry about building trust are not new. They are grounded in the literature and in personal experience. There are excellent publications that provide much greater detail than I have in this book (Covello et al. 1988; Committee on Risk Perception and Communication 1989; Persensky et al. 2004; Stern and Fineberg 1996). The major difference between my five suggestions is that I start with the assumption of a recent event:

1. Do not shoot the messenger. When people come to meetings with tough questions, are worried, angry, and even accusatory, do listen, do respond, and do not be dismissive. Reacting defensively that it could not happen in the USA at nuclear power plants or waste management facilities eliminates the reality that they have witnessed on television. Fear needs to be acknowledged, not dismissed as irrational.

2. If the audience wants to understand what happened in Japan, what have been the consequences so far, if you can, then carefully go through the sequence of events detailing what worked and did not work. If you cannot, try to arrange for someone who can for another meeting.

3. Even if they do not want to engage in step 2, people want the responsible parties in the USA to acknowledge their fear but demonstrate the organizations' ability to protect them and that safety is the highest priority. Telling the public that nuclear materials are important in medicine, food security, and for other purposes is not a bad idea, but not necessary and could even lead some participants to feel that the industry is excusing what happened in Japan, Chernobyl, or TMI as some sort of required trade-off. Also, telling the public that coal is environmental public enemy number 1 is a risk, especially among older participants who may not believe in global climate change and do remember that coal heated their homes. It is much more important to address the multiple backup systems in nuclear faculties, the barriers that prevent exposures, and the other lessons learned by nuclear scientists and engineers as a result of TMI, Chernobyl, and now Fukushima. The fact they every plant has a mock control room that operators use to practice their response to possible events, I have found to be reassuring rather than unnerving. The fact that every facility is involved in periodic mock security events is a reality that some say scares people. I find that it does the opposite, although not with all five groups. Not telling audiences about risk management programs because they may be used by opponents prevents supporters and the stealth groups from obtaining the information in a public setting.

4. Hearing that responsible parties are competent, are communicating by listening as well as talking, and sharing key values with responsible parties are all necessary to build trust post-Fukushima, but not sufficient. The public wants to hear promises about definitive steps to improve safety, efficiency, and other metrics of competence; they want to know what communications will keep them

or their representatives in the loop; and they want to be sure that the safety value is not going to be sacrificed. As long as public airing does not compromise security, then promises need to be made.

5. The last part of my five-part prescription is to follow-through with promised actions. Trying to change public opinion with words not matched by deeds is counterproductive. For example, I recommended to a government agency (nothing to do with nuclear materials) that they have an answering service to respond to public questions. They followed part of my suggestion, which was to install the answering service. They did not follow the other part of the suggestion, which was to have someone on call to respond to the questions, or if that was not possible to tell callers that their question would be answered in 24 h. By only following part of the suggestion, the organization undermined the credibility of their program in the eyes of angry callers who were feeling used (the organization explained to me later that they were using the phone system to determine what the public wanted to know). In retrospect, the organization would have been far better off doing nothing than what they did.

Different responses should be expected from the five major groups. The antinuclear group will fight to avoid hearing a positive tone and will be suspicious, at least initially, of assertions that safety is the highest priority, and the group members will be dubious of promises that involve substantial resource commitments. Any suggestion that a remediation schedule will slip will increase their lack of trust. It may seem counterproductive to spend a disproportionate amount of resources responding to opponents because most will remain opponents. Communications and trust building with opponents will be difficult (Keller et al. 2012). Softening of a strong negative position is a more reasonable objective.

The suggested post-Fukushima approach will also encounter a different kind of resistance from the uninterested group. They would fall at the bottom two of Arnstein's (1969) eight-step ladder of citizen participation. The challenge will be to engage them. Our surveys show that this group does not speak with fellows about the issues in this book, nor does it consult electronic or print media about the subject. If there is a mass media story, they may see it or hear it. But such messages are typically terse, slogans rather than substance. What will be heard by the disinterested group, unless there is a crisis, is at best a few slogans, which is clearly disconcerting.

The older cost-conscious, pronuclear, and fossil fuel group is more likely than its disinterested counterparts to read the print media. The five suggestions will resonate. However, if the nuclear industry delivers the message and includes an anti-fossil fuel (coal is environmental enemy number 1), this group may tune out the positive messages.

The message will resonate with the supporter and probably the stealth groups. The supporter group already trusts the responsible parties and shares similar values (see Dietz et al. 2005 for a discussion of the range of values). The challenge is how to build a bridge that will carry over to the broad range of issues that a long-term waste management program will face for many decades. I do not think that science and technology are the right bridge to build because the partners (government officials, community representatives, city planners) are separated from the DOE, contractors, and others in training and experience. My recommendation is land use, transportation, and other solvable problems (Burger et al. 2004; Connell and Pickett 2000; ICMA 1996; Wernstedt and Hersh 1997; Probst et al. 1997). These are issues that community representatives can contribute important insights. The DOE has site-specific advisory boards at key sites that have already participated in future use plans (Chap. 2; Greenberg et al. 2008). This bridge can be strengthened as issues such as moving defense waste and used fuel arise, siting and new transportation issues continue, how to adjust programs as federal budgets rise and fall, and where and how to build facilities that will house key stewardship data for future generations.

DOE's environmental management missions at the remaining 17 sites can be divided into four levels beginning with the easiest: decommissioning and decontamination of buildings and other structures and reuse of some sites; glassification and grout of tank wastes and other nonmobile hazards; preventing wastes from further migration into water bodies; and figuring out how to manage long half-life hazards. The DOE will benefit by hearing local suggestions about at least the first three of these four.

How scientific uncertainty is dealt with is important in interacting with the public. On the one hand, some believe that uncertain information confuses the public and leads them to embrace the worst possible scenarios. Yet not acknowledging uncertainty means that opponents can introduce uncertainty. It may be that reducing uncertainty about managing a waste material is more important to community group representatives than spending resources to complete a low-risk project. I believe that DOE can provide public representatives with plausible management options, such as spending half of a shrinking budget on removing the last 1 % of contaminants from an underground tank versus spending the same amount of money on multiple other projects at the same location. I am not so naive as to believe that the DOE's signed agreements that commit them to specific actions by specific dates will be renegotiated. Yet, I also know from experience that parties can readjust priorities if they can agree about them. Resets of this magnitude can only occur if the partners trust that the DOE, NRC, EPA, states, and others are presenting options in good faith and the agencies have the power to persuade others above them.

I believe that DOE's community programs built around SSABs allow for this possibility (Greenberg et al. 2008). I know of no similar organizational mechanism that would even serve as a platform regarding nuclear power plants and wastes stored at them. I acknowledge that commercial wastes were not supposed to be stored at the power plant sites. But they are, and power plant sites face potential issues regarding transportation, evacuation, and expansion at existing sites that merit a thoughtful institutional response somewhat akin to what is done for shareholders at some of the bigger DOE sites.

Risk preferences and perceptions are typically referred to as "soft" issues, which I think is the wrong message to community representatives and their staff.

Whatever the label, these issue preferences and perceptions do polarize people, cause fear, and undermine confidence in government and business that could linger and undermine genuine efforts to reduce risk. This means that elements of DOE, NRC, and EPA that need to rebuild trust in the short term and maintain it for generations must be strengthened because proactive actions are more successful than reactive ones (Fischhoff 1995). I urge the government agencies to establish continuous education programs for managers and others who are likely to interact with the public and their representatives. This would be facilitated by the appointment of advisory boards with expertise in psychology, communication, sociology, planning and economics, and providing information for media (Greenberg et al. 2009; Environmental Health Center 2001; Lofstedt 2005).

Building trust post-Fukushima is complicated by increasingly adversarial politics. Sadly, it seems improbable that our senior elected officials will be able to agree on a national energy strategy and plan, which leaves the public in the position of continuing to be bombarded with inconsistent and argumentative assertions. Accordingly, responsible parties need to stop hiding behind the shield that they are charged with hard science and technology, and instead they need to invest not only in technology and education to improve risk management but also in outreach programs that will be needed for generations. After decades of working on siting of nuclear facilities and risk management, I state without equivocation that there is no way of finding cleverer ways of explaining risk that will convince the vast majority of Americans that the risks scientists worry about are the same risks that they should worry about. A well-conceived and ongoing outreach effort is required to build technical competence, forthrightly communicate it, and demonstrate values that are consistent with or at least acceptable to the public.

References

Arnstein S (1969) A ladder of citizen participation. J Am Inst Plann 35:216–224

Blue Ribbon Commission on America's Nuclear Future (2012) Final Updated Report. http://brc. gov Accessed March 21, 2012

Burger J, Powers C, Greenberg M, Gochfeld M (2004) The role of risk and future land use in cleanup decisions at the Department of Energy. Risk Anal 24:1539–1549

Committee on Risk Perception and Communication, National Research Council, (1989) Improving risk communication. National Academy Press, Washington, DC

Connell J, Pickett D (2000) Land use controls on BRAC bases. ICMA, Washington DC

Covello V, Sandman P, Slovic P (1988) Risk communication, risk statistics, and risk comparisons: a manual for public managers. Chemical Manufacturers Association, Washington DC

Dietz T, Fitzgerald A, Schwom R (2005) Environmental values. Ann Rev Environ Resour 30:335–372

Environmental Health Center and National Safety Council (2001) A reporter's guide to Yucca mountain. Washington, DC

Fischhoff B (1995) Risk perception and communication unplugged: twenty years of research. Risk Anal 15:137–145

Greenberg M, Lowrie K, Hollander J, Burger J, Powers C, Gochfeld M (2008) Citizen board issues and local newspaper coverage risk, remediation, and environmental management. Remediation. Summer; 79–90

Greenberg M, Lowrie K, West B, Mayer H (2009) The reporter's handbook on nuclear materials, energy, and waste management. Vanderbilt University Press, Nashville, TN

Greenberg M, Truelove H (2011) Energy choices and perceived risks: is it just global warming and fear of a nuclear power plant accident? Risk Anal 31(5):819–831

Greenberg M, Mayer H, Powers C (2011) Public preferences for environmental management practices at DOE's nuclear waste sites. Remediation 21:117–131

Huizenga D (2012) WM symposia 2012 and FY 2013 budget overview. Paper copy received March 15, 2012

ICMA (1996) Cleaning up after the cold war: the role of local government in the environmental cleanup and reuse of federal facilities. ICMA, Washington DC

Jenkins-Smith H, Silva C, Nowlin M, deLozier G (2011) Reversing nuclear opposition: evolving public acceptance of a permanent nuclear waste disposal facility. Risk Anal 31:629–644

Keller C, Visschers V, Siegrist M (2012) Affective imagery and acceptance of replacing nuclear power plants. Risk Anal 32:464–477

Leiserowitz A, Maibach E, Light A (2009) Global warming's six Americas, Yale Project on Climate Change, http://www.americanprogress.org/issues/2009/05/6american.html. Accessed march 21, 2012

Lofstedt R (2005) Risk management in post-trust societies. Earthscan, London

Miller B, Sinclair J (2012) Risk perceptions in a resource community and communication implications: emotion, stigma, and identity. Risk Anal 32:483–495

Office of Environmental Management (2000) Paths to closure, status report. DOE/EM-0526

Persensky J, Browde S, Szabo A, Peterson L, Specht E, Wight E (2004) Effective risk communication, the nuclear regulatory commission's guideline for external risk communication, NUREG/BR-0308. USNRC, Washington, DC

Probst K, McGovern M, McCarthy K (1997) Long-term stewardship and the nuclear weapons complex. RFF, Washington DC

Stern P, Fineberg H (eds) (1996) Understanding risks: informing decisions in a democratic society. National Academy Press, Washington DC

U.S. Department of Energy (2012) Department of energy: Successes at the Recovery Act, http://energy.gov/sites/prod/files/RecoveryActSuccess-January 2012. Accessed March 21, 2012

Wernstedt K, Hersh R (1997) Land use and remedy selection; experience from the field – the Fort Ord Site. Resources for the future, Discussion Paper, Washington DC, 97–128

Index